Toward Affordable Systems

Portfolio Analysis and Management for Army Science and Technology Programs

Brian G. Chow, Richard Silberglitt, Scott Hiromoto

Prepared for the United States Army

ARROYO CENTER

The research described in this report was sponsored by the United States
Army under Contract No. W74V8H-06-C-0001.

Library of Congress Cataloging-in-Publication Data is available for
this publication.

ISBN 978-0-8330-4682-6

The RAND Corporation is a nonprofit research organization providing
objective analysis and effective solutions that address the challenges
facing the public and private sectors around the world. RAND's
publications do not necessarily reflect the opinions of its research clients
and sponsors.

RAND® is a registered trademark.

Published 2009 by the RAND Corporation
1776 Main Street, P.O. Box 2138, Santa Monica, CA 90407-2138
1200 South Hayes Street, Arlington, VA 22202-5050
4570 Fifth Avenue, Suite 600, Pittsburgh, PA 15213-2665
RAND URL: http://www.rand.org/
To order RAND documents or to obtain additional information, contact
Distribution Services: Telephone: (310) 451-7002;
Fax: (310) 451-6915; Email: order@rand.org

Preface

The objective of this project was the application of the RAND Corporation's Portfolio Management (PortMan) approach to U.S. Army applied research and advanced technology development. In our analysis, we applied and expanded PortMan to help the Army achieve a better match between its capability gaps and the products of its science and technology (S&T) programs at an affordable cost.

In the course of the study, we broadened our aim to develop a method and model not only suitable for the selection and management of the Army's S&T portfolio, but also applicable to projects at later stages of development and to other individual services' projects, or to all military projects. This monograph describes the method and model, and sources and procedures used to estimate the input parameters. It also demonstrates example applications for assisting decisionmaking on various important S&T issues. However, it should be emphasized that this study focused on methodology development. To demonstrate the methodology, we took a retrospective look at how well a small subset of 2005 Army Technology Objective (ATO) projects could meet a hypothetical set of gaps, which we assumed occurred in every force operating capability (FOC) requirement and sub-requirement. In other words, the study did not use data on real capability gaps and provided no information on how well the Army S&T portfolio meets the actual gaps. Moreover, since the ATOs were evaluated against hypothetical gaps, one should not draw any conclusions about the merits or drawbacks of any specific S&T project discussed in this study in meeting real Army capability gaps.

This monograph should be of interest to S&T and acquisition managers who are responsible for portfolio management of programs; engineers in research, development, test, and evaluation programs; and those who are interested in the optimal allocation of funds among different programs and/or developmental stages to yield the lowest total lifecycle cost in meeting future capability gaps.

This research was sponsored by Stephen Bagby, Deputy Assistant Secretary of the Army (Cost and Economic Analysis), Office of Assistant Secretary of the Army (Financial Management and Comptroller) and conducted within RAND Arroyo Center's Force Development and Technology Program. RAND Arroyo Center, part of the

RAND Corporation, is a federally funded research and development center sponsored by the United States Army.

The Project Unique Identification Code (PUIC) for the project that produced this document is SAFMR08810.

For more information on RAND Arroyo Center, contact the Director of Operations (telephone 310-393-0411, extension 6419; FAX 310-451-6952; email Marcy_Agmon@rand.org), or visit Arroyo's Web site at http://www.rand.org/ard/.

Contents

Figures

Tables

Summary

The Army Modernization Plan states:

> Historically, almost 70 percent of a system's total costs were incurred once the system had entered the operations and sustainment phase. As a result, decisions made during design and development place an enormous impact on the overall cost, sustainment, and readiness of items introduced into the Army inventory (U.S. Army, 2006, pp. 26–27).

This plan calls for the consideration of the lifecycle cost of a system during its design and development stage. Studies have shown that on average, 85 percent of life-cycle cost decisions have been made by the end of technology development. This fact raises the concern that, after the S&T phase, one cannot change the basic design to significantly reduce a system's lifecycle cost, further emphasizing the need to consider lifecycle cost at an early stage.

The current method used by the Army Deputy Assistant Secretary for Research and Technology to evaluate ATOs does not include lifecycle cost. Thus, the purpose of this study is to develop and demonstrate a method and a model that allow the consideration of lifecycle cost at an early stage of a system's development, so that timely corrective actions can be taken if lifecycle cost moves above an acceptable level. In addition to the incorporation of lifecycle cost, the method is aimed at alleviating the problem of summing or comparing "apples and oranges" that often occurs when ranking S&T projects. Typical schemes used in the Army S&T community for selecting or ranking S&T (or other) projects sum an individual project's contributions to meeting several different future capability requirements. This traditional approach to ranking runs the risk of having projects scored high and selected for continued funding, even though their high scores may all come from meeting the same FOC requirements. A portfolio of such highly ranked projects can leave the requirements unmet for the rest of the equally important FOCs.

In addition to introducing a new method, the study also demonstrates how input parameters, including cost components, can be estimated. However, the Army S&T and acquisition community will very likely have much better data and can more easily and accurately make cost estimates. Readers who do not desire to use our cost estimation approaches may still find our analytic framework necessary in dealing with the aforementioned shortfalls currently encountered.

This study takes advantage of the RAND PortMan research and development (R&D) portfolio analysis and management method, which has been under development and improvement since 2002. In addition to consideration of lifecycle cost and insistence on meeting all FOC requirements individually, we have developed a linear programming model for selecting an optimized portfolio of ATOs that can meet all the FOC requirements at the lowest total remaining lifecycle cost.[1]

Method and Model

The genesis of this method is the RAND PortMan R&D portfolio analysis and management method, which is based on the determination of an expected value (EV) of R&D projects and portfolios. While the previous versions of PortMan, developed for the U.S. Department of Energy and used for Navy applied research programs, allowed detailed comparisons of R&D projects and provided guidance for the portfolio manager relevant to how to manage the R&D portfolio to maximize EV, investment strategy decisions were based on EV alone, with cost considerations based on the EV analysis introduced *ex post facto*.

In this monograph, we describe a variant of PortMan that uses a linear programming model to select a portfolio that consists of ATOs, the highest priority Army S&T projects. This portfolio satisfies the Army's FOC requirements designated for this group of ATOs to meet, while yielding the lowest total remaining lifecycle cost for all the systems developed from the selected ATOs. Our model includes two classes of constraints. First, all individual FOC requirements must be fully met. Second, the total remaining S&T budget[2] for the selected ATOs must not exceed a given budgeted amount that the Army can afford or is willing to pay. As a key application of this model, the budgeted amount was varied to study the relationship between the "affordable" total remaining S&T budget and the "affordable" total remaining lifecycle cost of the systems deployed. The result is that, in some cases, the Army may want to increase the total remaining S&T budget in order to significantly lower the total remaining lifecycle cost. Because the total remaining S&T budget is only a small portion of the total

---

[1]   Total remaining lifecycle cost is (1) the future lifecycle cost that still has to be paid in order to complete the selected ATOs and to develop and demonstrate the new systems derived from the ATOs and (2) the cost difference between (a) acquiring units of these new systems over a 20-year acquisition period and in operating and maintaining them over their lifetimes and (b) those of the legacy systems. The total does not include past lifecycle cost, which is already spent and should not enter into future decisions.

[2]   Total remaining S&T budget is the future S&T budget required to complete the S&T projects (ATOs) selected for the portfolio. It is part of the total remaining lifecycle cost.

remaining lifecycle cost, some S&T projects, although they have high remaining S&T cost, can yield systems that are significantly cheaper to develop, demonstrate, acquire, operate, and maintain—thus lowering the total remaining lifecycle cost far more than the increase in the total remaining S&T budget required to fund these projects.

Inputs to the Linear Programming Model

The linear programming model used here requires two inputs: (1) the EV of the ATOs and (2) the remaining lifecycle cost of the systems that will be developed from the ATOs. The data that we used to estimate the EVs of the ATOs were derived from the 2006 Defense Technology Objectives (DTOs), which were the latest available during the time of our study. DTOs are updated regularly on the Web by the Director of Defense Research and Engineering. All of the ATOs we chose for this example application of our method and model were also DTOs. We selected ATOs that are also DTOs for this study because more data are available on DTOs.

We have developed an approach called "gap space coverage" to make EV estimates of how well each ATO could meet FOC sub-requirements defined by the U.S. Army's Training and Doctrine Command. This approach is based on the multiplication of three factors: (1) how many situations encountered by warfighters to which the system derived from the ATO can make a contribution, (2) the size of the gap space in an FOC that the system can help fill, and (3) the size of the contribution that the system can make to filling the gaps.[3]

To estimate remaining lifecycle cost, we started with the remaining S&T cost from the DTO data sheets. We then estimated the System Development and Demonstration cost from the same costs of comparable existing systems. Finally, we added the marginal cost over the legacy system, which is the cost of the new system minus that of the existing system it replaces. This cost has three components: (1) acquisition, (2) upgrade, and (3) operating and maintenance.

Applications to S&T Portfolio Management

For the demonstration of our method and model, we chose the 29 Army ATOs that are DTOs (as mentioned above). Since these ATOs are only a small subset of the 172

---

[3] However, it should be emphasized that this study focused on methodology development. To demonstrate the methodology, we took a retrospective look at how well a small subset of the 2005 ATO projects could meet a hypothetical set of gaps, which we assumed occurred in every FOC requirement and sub-requirement. In other words, the study did not use data on real capability gaps and provided no information on how well the Army S&T portfolio meets the actual gaps. Moreover, since the ATOs were evaluated against hypothetical gaps, one should not draw any conclusions about the merits or drawbacks of any specific S&T project discussed in this study in meeting real Army capability gaps.

existing ATOs, one should not expect this subset to meet 100 percent of each of the FOC requirements. Based on a rough linear approximation, we assume that these 29 ATOs should be required to meet their fair share of 17 percent (i.e., 29 ÷ 172) of the requirements (required expected value [REV]). In fact, as shown by the blue bars in Figure S.1, these 29 ATOs together have the potential to meet on average 46 percent of the FOC requirements, well exceeding their 17 percent "fair share," which would have been achieved at 37 percent (i.e., 17 ÷ 46) of the actually achievable 46 percent. In this study, the 37 percent is rounded to 40 percent, which is then used as the FOC requirement level that any portfolio of ATOs selected from the 29 ATOs is asked to meet. Additionally, if the systems derived from all 29 ATOs were fielded, they would produce extra capabilities above the capability requirements, as indicated by the red bars in Figure S.1. The difference between the blue and red bars indicates the depth of the ATO pool.

To rank the existing ATOs according to their importance in meeting FOC requirements, we defined the blue bars in Figure S.1 as the achievable EVs that all 29 ATOs together can attain. We then applied our linear programming model to select a subset of these 29 ATOs that meets the REVs, so that the total remaining lifecycle cost is minimized. Results are shown in Figure S.2, where the reference requirement (RR) is 40 percent of achievable EVs. Because the future Army budget for total remaining lifecycle cost and the future strategic environment are uncertain, the requirement may turn out to be higher or lower than the reference. For example, slower economic

Figure S.1
Achievable and Required Expected Value for 29 ATOs

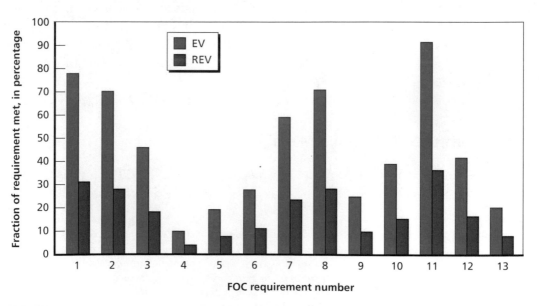

Figure S.2
ATOs That Meet Various FOC Requirements at the Lowest Total Remaining Lifecycle Cost

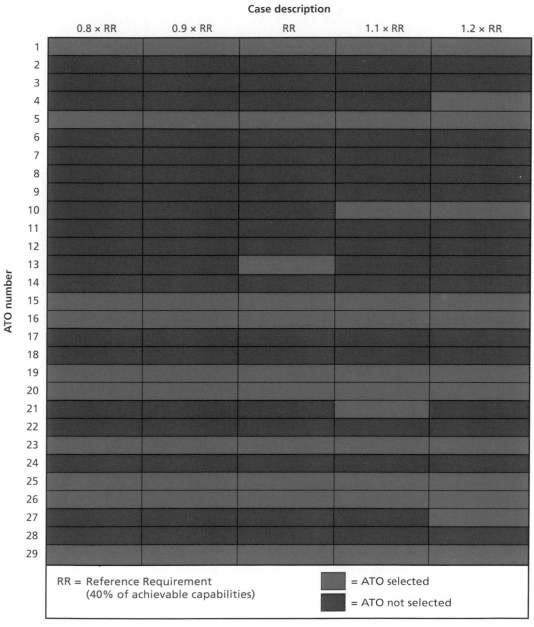

growth can result in a lower Army budget and force the Army to lower its requirement. On the other hand, a new strategic environment following a major strategic event, such as post–9/11, can lead to additional requirements. ATOs that are selected to yield the lowest total remaining lifecycle cost regardless of future requirement levels should be ranked high and protected. Since 1988, the annual defense budget has fluctuated 29 percent above and 21 percent below the average. Because we use a weapon deployment period of 20 years, the uncertainty in a budget and a requirement averaging over a 20-year period is likely to be smaller than the annual uncertainty, and we assume it to be 20 percent. At the reference requirement of 40 percent, this represents a potential range of 32 percent up to 48 percent. The question can therefore be stated as follows:

> Given a reference requirement of 40 percent, when the FOC requirements are set at 0.8 (i.e., 32 percent), 0.9 (36 percent), 1 (40 percent), 1.1 (44 percent), and 1.2 (48 percent) of the reference requirement, which ATOs are selected most often for these five cases in order to yield the least total remaining lifecycle cost?[4]

Those ATOs selected most often are most important because no matter what the future will be, they are needed to result in the lowest total remaining lifecycle cost.

In Figure S.2, a green cell indicates that an ATO was selected under the particular FOC requirement. A red cell indicates that the ATO was not selected. ATOs 1, 5, 15, 16, 19, 20, 23, 25, 26, and 29 are always selected under every requirement level. They are the highest ranking ATOs and should be best protected in the event of a budget cut. ATO 10 is selected twice, while ATOs 4, 13, 21, and 27 are selected once. These ATOs are still needed in some future possible requirement levels. In order to meet the requirements at all levels at the lowest total remaining lifecycle cost, as the second- and third-ranked ATOs they should be protected as well. The rest of the 29 ATOs are never selected. In this study, all ATOs are assumed to be successfully completed if they are funded until completion. This assumption implies that there is no need for any backup ATOs, since the selected ATOs will not fail. In our ongoing work, all ATOs are assumed to have a possibility of failure, and we examine the possibility of ATOs that are never selected here serving as backup in the event that the selected ones fail.

We have conducted two sensitivity tests. The first one looks at a future in which the Army requirements are more uncertain. The model shows which additional ATOs should be funded to completion in order to be able to meet a wider range of future requirements. The second one studies whether rougher-grain cost data (grouping our costs in dollars into seven levels) would lead to different model results. We found that the project selection is identical to that in Figure S.2. This suggests the usefulness of a much quicker and easier Delphi method[5] for cost estimation.

---

[4]   The reference requirement refers to the 13 REVs (red bars) in Figure S.1. At 0.8RR, we mean that all 13 individual requirements are reduced by 20 percent or set at 80 percent of the reference REVs.

[5]   The *Delphi method* uses a panel of experts to answer questionnaires or to grade parameters in two or more rounds. After each round, an anonymous summary of the experts' answers from the previous round is provided to all participants, who are encouraged to revise their earlier individual answers in view of the group's answers.

Our model demonstrates that the total remaining lifecycle cost required to meet marginal increases in FOC requirements is non-linear. The blue line in Figure S.3 shows the marginal cost to meet FOC requirements at four 10-percent increments. It takes $2.1 billion to meet 0.8RR (i.e., the first 80 percent of the reference requirement). However, the cost allocated to meet 0.8RR also meets the next 10 percentage points of the requirement (0.9RR), representing a marginal cost of zero dollars. As the requirement increases at equal 10-percent increments from 0.9RR to RR to 1.1RR and 1.2RR, marginal increases in total remaining lifecycle cost are $53 million, $780 million, and $830 million, respectively—in a non-linear fashion.

This non-linearity in meeting the four 10-percent increments from 0.8RR to 1.2RR is due to the complicated interplay of two factors. First, all else being equal, the model selects ATOs that can develop systems to meet the requirements most cheaply. Second, adding an additional ATO can sometimes meet requirements in larger chunks than the 10-percent increment. For example, the group of ATOs selected by the model to fulfill the first 80 percent of the reference requirement also fulfills the first 90 percent of the requirement without needing any additional ATO or funding.

Acquisition planners need to pay attention to this non-linearity of cost when they are asked to estimate the change in total remaining lifecycle cost to meet increased or decreased FOC requirements. A good decision hinges on knowing not only the change of requirement, but also the change in lifecycle cost to meet the changed requirement.

Figure S.3
Remaining Lifecycle Cost for Four 10-Percent Increases in Requirements

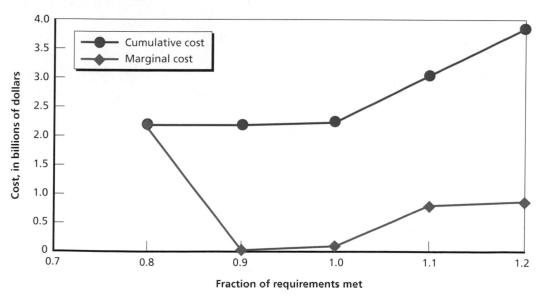

RAND *MG761-S.3*

For example, if the additional lifecycle cost to meet a large increase in FOC requirements is low, acquisition planners are more justified in seeking the added funds. On the other hand, if only a small savings in lifecycle cost would result from a large reduction in FOC requirements, the acquisition planners also need to make this situation known to the Army's senior leadership so that the same message can be conveyed to the Secretary of Defense and the legislators in order to avoid a large reduction in the requirements.

Our model also allows examination of the important relationship between the total remaining S&T budget and the total remaining lifecycle cost. Figure S.4 shows this relationship, where the total remaining S&T budget is that required to complete the ATO projects in the portfolio selected to meet all individual FOC requirements (the red bars in Figure S.1) at the lowest total remaining lifecycle cost. In this case, the model selects 11 of the 29 ATOs, with a total remaining S&T budget of $206 million and total remaining lifecycle cost of $2.2 billion. Figure S.4 shows that the total remaining lifecycle cost increases sharply if the total remaining S&T budget is reduced from $206 million only slightly. The cause for this sharp rise is that a smaller total remaining S&T budget forces one to choose cheaper ATOs, although these ATOs would develop systems that are a lot more expensive to field and operate, and they consequently increase the total remaining lifecycle cost sharply. For example, when the total remaining S&T budget is limited to $159 million, the total remaining lifecycle cost rises from $2.2 billion to $9.2 billion. In other words, in our demonstration case, saving $47 million in remaining S&T funds in the near term would cause the Army to pay $7 billion more in total remaining lifecycle cost in the long term. This drastic outcome results from the fact that the total remaining S&T budget is typically a very small fraction of the total remaining lifecycle cost. Our model demonstrates the existence of an optimal level of the total remaining S&T budget, and the importance of maintaining the budget at this optimal level. Such a curve (as in Figure S.4) would provide the Army with an important guidance in the proper allocation of funds between total remaining S&T activities and other expenditures, such as procurement of weapon systems.

Above we illustrated only some of the applications of our method and model. The following is a much more complete list:

- Determination of the extent to which the FOC requirements would be met if all existing ATOs were completed and their systems fielded. This allows the Army to trace the impact of ATOs on FOCs.
- Identification and introduction of new ATOs for which existing ATOs leave gaps in meeting FOC requirements.
- Determination of the subset of existing ATOs that can meet all individual FOC requirements at the lowest total remaining lifecycle cost. This determination is of particular interest to the Assistant Secretary of the Army for Acquisition, Logis-

Figure S.4
Existence of an Optimal S&T Budget

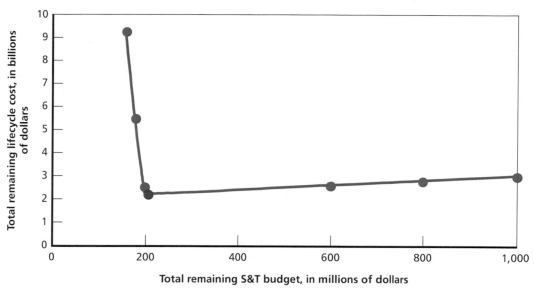

RAND *MG761-S.4*

tics, and Technology and the Army Vice Chief of Staff, who co-chair the S&T advisory group that provides the final approval of S&T programs.

- Determination of the extent to which each of the individual FOC requirements is exceeded. This provides a safety margin beyond that already embedded in the requirement. The Army—in particular its Training and Doctrine Command; the Assistant Deputy Chief of Staff for Programs (Force Development); and the Army's Research, Development and Engineering Command—need to know in which individual FOC requirements these extra safety margins will occur, because these are the requirements for which the Army can have extra capabilities without the need to develop additional costly ATOs and systems.

- Determination of the optimal distribution of funds across FOC requirements. We also found that distributing S&T funds and/or acquisition and operating funds evenly among individual FOC requirements is unlikely to be the least expensive way to meet FOC requirements.

- Determination of which set of ATOs should be ranked high and be protected from budgetary cuts. This set can yield the lowest total remaining lifecycle cost over a range of uncertain future requirements.

- Identification of cost and performance factors that prevent specific ATOs from being ranked higher in importance. Understanding these factors is useful to individual ATO program managers as they develop cost and performance targets.

The model determines the specific cost and contribution that an ATO needs to attain to move to a higher rank.

- Determination of the optimal level of the total remaining S&T budget to meet all of the individual FOC requirements at the lowest total remaining lifecycle cost. The non-linearity in the relationship between the total remaining S&T budget and the total remaining lifecycle cost makes this information critically important as the Army's leadership balances overall Army resource demands. We have found in our example demonstration that even a small cut in the total remaining S&T budget can lead to lifecycle costs in later years that are an order of magnitude or two larger than that of the cut. Also, in our example, a modest increase in the Army total remaining S&T budget can lead to a much larger savings in total remaining lifecycle cost.

Recommendations

During this study, we developed a method and model that factors in lifecycle costs and meets the conditions of all individual FOC requirements. We also demonstrated how this method and model can be applied to designing, funding, and managing the Army's S&T portfolio so as to meet all the Army's FOC requirements at the lowest total remaining lifecycle cost. The next step is to apply the method and model to real-life cases. We recommend three action items for such applications:

1. The Army S&T community sets up a pilot program between the Warfighting Technical Council and the offices in charge of Army Technology Objectives to test the practicality and usefulness of our iterative procedure for better estimating an ATO's contributions to individual FOC requirements and its derived system's lifecycle cost. This pilot program, if successful, will be an important step toward the goal of meeting future capabilities at the lowest cost.

2. The Army S&T community sets up a pilot program to carry out the method, model, and estimation procedures proposed in this monograph. The pilot program can be based on all of the existing ATOs and should include the eight applications suggested in the previous section, plus the use of the much simpler Delphi method for estimation of rough cost levels. The use of a Delphi method may lead to a wider adoption of our model by various agencies in the S&T and acquisition community.

3. Since the method and model developed here are applicable to the portfolio management of Army programs at other than the S&T stage, the Army may want to apply the approach to, for example, the selection and management of its programs in the System Development and Demonstration stage. Other services and the U.S. Department of Defense (DoD) may wish to test our approach as

Collate Color Text Section

Collate Color Text Section

well, since they also need to consider individual capability requirements and the lifecycle costs of their programs at an early developmental stage.

In an ongoing follow-on study, we are expanding our method to incorporate uncertainties in S&T project successes with the objective of making the model a better reflection of the uncertain future. The ongoing study examines the impacts of uncertainties on such issues as the selection of ATOs that can meet FOC requirements over a wide range of uncertain futures. In other words, we want to describe how to select a portfolio of ATOs that is robust against uncertainties.

We believe that not only the Army but also the other services and DoD as a whole need to consider lifecycle cost at an early stage in the development of weapon and other systems in order to allow adjustments, where necessary, to achieve affordable systems that meet all individual capability requirements.

Acknowledgments

We are grateful for the advice, guidance, and financial support of David Henningsen of the Office of the Deputy Assistant Secretary of the Army for Cost and Economics (DASA-CE), without which this study could not have been possible. We are also grateful to Myrna Kroh of U.S. Army G8-FDA, who brought RAND's PortMan method to the attention of the Army and provided substantial support for the development of the study plan. We are pleased to acknowledge and thank our peer reviewers, Walter Perry and Lance Sherry, who provided many insightful comments and suggestions that contributed significantly to improving the exposition and explanation of our method, model, and results. Finally, we thank Bruce Held, our program manager, for providing prompt and valuable comments during every stage of the project, from proposal to interim report to final monograph.

Abbreviations

AAE	Army acquisition executive
ACAT	Acquisition Category
ACTD	Advanced Concept Technology Demonstration
AFATDS	Advanced Field Artillery Tactical Data System
AoA	analysis of alternatives
ATD	advanced technology demonstration
ATFLIR	Advance Targeting Forward Looking Infrared
ATO	Army Technology Objective
BLOS	beyond line of sight
BSTC	Budgeted Total Remaining S&T Cost
C2	command and control
C2I	command, control, and intelligence
C4	command, control, communications, and computers
CBTDEV	combat developer
CDD	Capability Development Document
CJCS	Chairman of the Joint Chiefs of Staff
CPD	Capability Production Document
CR	Concept Refinement
CV	coverage score
DARPA	Defense Advanced Research Projects Agency

DASA	Deputy Assistant Secretary of the Army
DASA-CE	Deputy Assistant Secretary of the Army for Cost and Economics
DDR&E	Director of Defense Research and Engineering
DoD	Department of Defense
DR	discount rate
DTO	Defense Technology Objective
EV	expected value
FAAD	Forward Area Air Defense
FCS	Future Combat Systems
FOC	force operating capability
GIG	Global Information Grid
HIV	human immunodeficiency virus
HMMWV	high-mobility multipurpose wheeled vehicle
ICD	Initial Capabilities Document
IED	improvised explosive device
IOC	Initial Operating Capability
IOT&E	Initial Operational Test and Evaluation
ISR	intelligence, surveillance, and reconnaissance
JCIDS	Joint Capabilities Integration and Development System
JIIM	joint, interagency, intergovernmental, and multinational
JROC	Joint Requirements Oversight Council
JWSTP	Joint Warfighting Science and Technology Plan
LOS	line of sight
MAIS	major automated information system
MCS	Mounted Combat System
MDA	Milestone Decision Authority
MDAP	major defense acquisition program

MIC	marginal implementation cost
MOMC	marginal operating and maintenance cost
MPC	marginal procurement cost
MRD	materiel requirement document
MRLCC	marginal RLCC
MUC	marginal unit cost
MUGC	marginal upgrade cost
NLOS	non–line of sight
NPLEG	number of legacy systems procured
NPNEW ·	number of new systems procured
O&M	operating and maintenance
OASA	Office of the Assistant Secretary of the Army
OIF	Operation Iraqi Freedom
OMCLEG%	O&M Cost of Legacy System (in percentage)
OMCNEW%	O&M Cost of New System (in percentage)
ORD	operational requirement document
OS	operations and support
OSMIS	Operating and Support Management Information System
PD	Production and Deployment
PE	program element
PortMan	(RAND's) Portfolio (Analysis and) Management (Method)
PPR	Plans, Programs, and Resources
R&D	research and development
RDT&E	research, development, test, and evaluation
REV	required expected value
RLCC	remaining lifecycle cost
RR	reference requirement

RSTC	remaining S&T cost
S&T	Science and Technology
SDD	System Development and Demonstration
SDDC	System Development and Demonstration cost
SINCGARS	Single Channel Ground to Air Radio System
TD	Technology Development
TRAC	TRADOC Analysis Center
TRADOC	Training and Doctrine Command
UAV	unmanned aerial vehicle
UCLEG	unit cost of legacy system
UCNEW	unit cost of new system
WTC	Warfighting Technical Council

CHAPTER ONE

Introduction

It is important to start gaining an understanding of a system's lifecycle cost[1] as early in that system's development as possible. It has been mentioned by acquisition officials and others that most of the lifecycle cost of a military system is determined during the design and development phase. In other words, if one only considers the lifecycle cost after this phase, it might be too late to adjust the system design to make fielded units affordable. Studies have shown that on average, 85 percent of lifecycle cost decisions have been made by the end of the technology development (Andrews, undated).[2] Since our study focuses on U.S. Army systems, it is of interest to note that Donald Damstetter, Deputy Assistant Secretary of the Army for Plans, Programs and Resources, Office of the Assistant Secretary of the Army (Acquisition, Logistics and Technology), cited and agreed with the above statement about the 85 percent of lifecycle cost decisions (Damstetter, 2004). Most important, the Army Modernization Plan stated:

> Historically, almost 70 percent of a system's total costs were incurred once the system had entered the operations and sustainment phase. As a result, decisions made during design and development place an enormous impact on the overall cost, sustainment, and readiness of items introduced into the Army inventory (U.S. Army, 2006, pp. 26–27).

This statement suggests that consideration of system lifecycle cost impacts should start as early as practicable. This is especially the case given the concern that making basic design changes to a system gets more impractical and expensive as development matures. This is as true for changes made in reaction to new system requirements as it is for those needed to reduce lifecycle cost.

---

[1] The lifecycle cost of a system is the cradle-to-grave cost from its research and development, to system development and demonstration, to acquisition and fielding, to operation and maintenance, to disposal.

[2] Moreover, a chart from the Air Force suggests that 90 percent of lifecycle costs are predetermined by the end of phase 6.3 (Advanced Technology Development) of the Science and Technology (S&T) program. That chart, from the Air Force Research Laboratory Material Directorate, appears in Moulder, undated. Regarding this paper by Moulder, although it is undated it can be deduced from its references that it was written after March 2000.

Recognizing the need to begin assessments of lifecycle cost early, the Department of Defense's (DoD) new acquisition policy, adopted in 2003, stipulates that an analysis of alternatives (AoA) be conducted for major systems when they enter into the Technology Development (TD) phase.[3] An AoA includes an estimate of lifecycle cost.

Since the Army wants to consider lifecycle cost during the design and development stage of its weapon and other systems and the DoD demands this early consideration, the Army, as well as other services, needs lifecycle cost to be incorporated into the analysis and management of projects in the S&T phase, as we will elaborate on further in Chapter Two.

Study Objectives

The Army and other services currently do not have a method for incorporating lifecycle cost considerations during the S&T phase of development. The purpose of our study is, therefore, to develop and demonstrate a practicable method for considering lifecycle cost at an early stage of a system's development. In this study, we applied our method to projects in the Army S&T programs, both to select an optimal S&T project portfolio at the start and to help inform portfolio managerial decisions, such as which projects to keep or terminate—especially when faced by a cut in the S&T budget—and what the optimal S&T funding level should be in relation to the much higher funding level for demonstrating, acquiring, operating, and maintaining Army systems.

In addition to capturing lifecycle cost, our method aims to rationalize the comparison and ranking of S&T projects. Typically selecting or ranking S&T (or other) projects is done by summing individual project contributions against various future capability requirements. For example, the Army has developed about 200 Army Technology Objectives (ATOs), the highest priority S&T projects, as solutions to meeting 11 force operating capabilities (FOCs).[4] The priority assigned to an ATO is typically determined by the sum of its contributions across the FOCs. Unfortunately, this approach creates the possibility that not all FOCs will be adequately addressed by a portfolio of high-ranking ATOs. It is possible that all the higher-priority ATOs address

---

[3] We will show in Chapter Two that the technology development stage discussed here should include at least the 6.3 programs in the S&T phase.

[4] In this study, we focus on the ATOs, which the Army considers "the most important S&T programs" (U.S. Army, 2005a, p. 265).

> To focus and stabilize the 6.2 and 6.3 programs, practice management by objective, and provide feedback to Army scientists regarding their productivity, the Army designates selected S&T efforts as Army Technology Objectives (ATOs). . . . Not every worthwhile funded 6.2 and 6.3 technology program is cited as an ATO. Since ATOs are part of a rigorous process to deliver technology within a scheduled time frame based on need, they are by their nature describing technologies that are better understood (U.S. Army, 2005d, p. I-8).

the same FOCs, so that some FOC requirements are over-met while others are unmet.[5] The method presented in this monograph was developed to select projects and ensure that they together will meet the requirements of all individual FOCs.[6]

In addition to introducing a new method, the study also demonstrates how input parameters, including cost components, can be estimated. In addition, the Army S&T and acquisition community likely has much better data available to it and thus may be able to make cost estimates more easily and more accurately than those presented here. Consequently, readers who do not want to use our cost estimation approaches may still find our analytic framework useful.

This study takes advantage of the RAND Portfolio Management (PortMan) research and development (R&D) portfolio analysis and management method, which has been under development and improvement since 2002 (Silberglitt and Sherry, 2002).[7] In addition to the two aforementioned objectives (incorporation of lifecycle cost into consideration and insistence on meeting all FOC requirements individually), we have developed a linear programming model for selecting an optimized portfolio of ATOs that can meet all the FOC requirements at the lowest total remaining lifecycle cost.[8]

Note that the numbers in this report are not rounded to their significant figures, which will allow other analysts to check our method and replicate our calculations more conveniently.

Problem Statement

As discussed above, *the current method for portfolio analysis of Army S&T programs does not include lifecycle costs.* Without lifecycle cost factored into the portfolio analysis, many key questions cannot be addressed. For example, the Army may miss the opportunity to select some ATOs that can lower the cost to field and operate the systems in

---

[5] This point can be easily understood with an extreme example. Let us assume that all the ATOs that are selected for continued funding, as opposed to termination, because of high scores for FOC 1. In contrast, the other ATOs are terminated because they have lower total scores, although they do contribute to other FOCs. In this extreme case, the selected portfolio of ATOs will lead to a future in which only FOC 1 will be met, with all other FOC requirements unmet. In other words, this portfolio has created too much capability for FOC 1 and no capability for others. Clearly, this is a very poor selection of ATOs for continued funding and is a reflection of the pitfall of ranking ATO attractiveness by totaling their contributions to different FOCs.

[6] As described in Chapter Two, we achieve this objective by adding the contribution of each project to each FOC, keeping the sum for each FOC separate, but making sure that the sum at least meets that FOC requirement.

[7] See Chapter Three below for details.

[8] Total remaining lifecycle cost is (1) the future lifecycle cost that still has to be paid in order to complete the selected ATOs and to develop and demonstrate the new systems derived from the ATOs and (2) the cost difference between (a) acquiring units of these new systems over a 20-year acquisition period and in operating and maintaining them over their lifetimes and (b) those of the legacy systems. The total does not include past lifecycle cost, which is already spent and should not enter into future decisions.

the future, although these ATOs may require a larger up-front S&T budget to complete. More generally, the Army does not know how increasing or decreasing the total remaining S&T budget[9] can affect the cost of fielding and operating the systems in the future, which accounts for the lion's share of the lifecycle cost. Therefore, the Army is unable to identify the optimal S&T budget relative to the systems' fielding and operating budget. Moreover, *the ranking of ATO attractiveness is based on the summation of an ATO's contribution to different FOCs*, raising the possibility that the portfolio of ATOs selected for continued funding until completion will not meet some FOC requirements.

The aim of this study is to develop a method that can address the two problems described and italicized in the previous paragraph.

Report Structure

Chapter Two is a review of DoD's current acquisition policy, focusing on requirements for early consideration of lifecycle cost during system development, which is the part most relevant to this Army study. Chapter Three describes the evolution of RAND's PortMan portfolio analysis and management method. It also develops a linear programming model that is used to factor individual capability requirements and lifecycle costs into the latest version of the PortMan method. Chapter Four discusses the sources and methods for estimating the input parameters to our model. Chapter Five explains various ways in which the model outputs can be used to inform decisions concerning the selection and management of S&T projects. Finally, Chapter Six is a summary of the findings and recommendations concerning how this method and model can be used to aid decisionmaking on a variety of issues. Additional explanatory material is presented in the appendixes. Appendix A describes the primary criteria for classifying an Army Acquisition Category (ACAT) and the role of a combat developer (CBTDEV), which pertains to a deeper understanding of acquisition policy, which is discussed in Chapter Two. Appendix B discusses how an ATO's contributions to individual FOC requirements are determined. Appendix C describes the methods we used to estimate input cost parameters. Appendix D describes the methods we used for extrapolating our cost estimation. Appendix E discusses a sensitivity analysis with requirement uncertainties in the plus and minus 50 percent range, as opposed to the 20 percent range used in the text. Appendix F lays the groundwork for a Delphi method to estimate remaining lifecycle cost for individual ATOs and their systems. The approach is much simpler and quicker than our detailed cost estimation. It can be used as a component in our proposed iterative procedure for arriving at better cost estimates.

---

[9] Total remaining S&T budget is the future S&T budget required to complete the S&T projects (ATOs) selected for the portfolio. It is part of the total remaining lifecycle cost.

CHAPTER TWO

The Current DoD Acquisition Policy

This chapter begins with a description of the general process used by DoD to identify and deliver required materiel capabilities. This is followed by an explanation of Army acquisition program categorization and how these different categories compare in terms of documentation and approval procedures at program milestones. Next, we discuss how DoD acquisition policy is applied to Army system acquisitions and under what circumstances an AoA is required. Finally, the feasibility and necessity of lifecycle cost considerations for ATOs are examined by drawing parallels with the required documentation and approval procedures for major systems.

Joint Capabilities Integration and Development System

The Chairman of the Joint Chiefs of Staff (CJCS) and the Joint Requirements Oversight Council (JROC) use the procedures established in the Joint Capabilities Integration and Development System (JCIDS) to identify, assess, and prioritize joint military capability needs (CJCS, 2007a, pp. 1–2). The JCIDS was created to support the statutory requirements of the JROC; to validate and prioritize joint warfighting requirements; to develop, acquire, and field the required systems; and to budget for the whole process, from concept refinement to operations and support of the deployed system. JCIDS has three key processes: the requirements process; the acquisition process; and the Planning, Programming, Budgeting, and Executive process. Because the acquisition process is most relevant to this study, the discussion focuses on it.

Army Acquisition Categories

Defense acquisition programs are classified into distinct ACATs that determine the level of the Milestone Decision Authority (MDA).[1]

---

[1] This section is based on U.S. Army, 2003, pp. 30–33. This 2003 Army document had taken into account the changes from U.S. Department of Defense, 2003a and 2003b.

- ACAT I programs are major defense acquisition programs (MDAPs) and have two subcategories: ACAT ID and ACAT IC. ACAT ID programs are MDAPs whose MDA is the Under Secretary of Defense for Acquisition, Technology and Logistics. ACAT IC programs are MDAPs whose MDA is the service acquisition executive. In this Army study, the service acquisition executive is the Army acquisition executive (AAE),[2] who reviews an acquisition program's progress and approves it at each milestone before it enters into the next acquisition phase, as shown in Figure 2.1.
- ACAT IA programs are major automated information systems (MAISs) or programs. ACAT IA has two subcategories: ACAT IAM and ACAT IAC. The former programs are MAISs whose MDA is the DoD Chief Information Officer, while the latter is the AAE.
- ACAT II programs are those programs that do not meet the criteria for an ACAT I program (i.e., ID and IC), but are major systems or are designated as ACAT II by the AAE.
- ACAT III programs are those that do not meet the criteria for ACAT I (i.e., ID and IC), ACAT IA, or ACAT II. The MDA is designated by the AAE. They are non-major systems, including command, control, communications, and computers and information technology.
- The Army eliminated the designation of ACAT IV programs in an Army Acquisition Policy memorandum issued on December 31, 2003.

Figure 2.1
The Defense Acquisition Management Framework

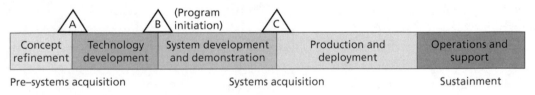

SOURCE: U.S. Department of Defense, 2003b, p. 2.
RAND *MG761-2.1*

MDA is the office that makes the decision at milestones whether the acquisition program can progress to the next stage.

[2]   The AAE is the Honorable Claude M. Bolton, Jr., the Assistant Secretary of the Army (Acquisition, Logistics and Technology).

Collate Color Text Section

Collate Color Text Section

DoD Acquisition Policy and AoA

On May 12, 2003, DoD issued an updated acquisition policy. Major acquisition programs are required to follow various milestones and approval procedures through five phases: Concept Refinement (CR), Technology Development (TD), System Development and Demonstration (SDD), Production and Deployment (PD), and Operations and Support (OS), as shown in Figure 2.1. Through the JCIDS process, as described in an earlier section of this chapter, the joint military capability needs or gaps are determined. The capability-based assessment (CBA) is the analysis part of the JCIDS process, and it has three major outputs (CJCS, 2007b, p. A-1):

- The functional area analysis (FAA), which describes the mission area being assessed.
- The functional needs analysis (FNA), which assesses how well the current or programmed force performs that mission. If the performance is not good or there is a need to improve performance, a military capability gap or need is identified.
- The functional solutions analysis (FSA), which analyzes and proposes possible solutions to shortcomings or to problems in attaining the needed capability.

The materials from these three outputs are used in the preparation of an AoA. At a milestone, MDA reviews the required documents and decides whether the program should proceed to the next phase. DoD stipulates that "the ICD [Initial Capabilities Document] and the AoA shall guide Concept Refinement." Moreover, "the results of the AoA shall provide the basis for the TDS [Technology Development Strategy], to be approved by the MDA at Milestone A for potential ACAT I and IA programs" (U.S. Department of Defense, 2003b, p. 6). In other words, an AoA is required and due at Milestone A for all ACAT I (i.e., ID and IC) and ACAT IA (i.e., IAM and IAC) programs. Moreover, AoAs are recommended for all others (Army Logistics Management College, 2005, p. 6). While the Training and Doctrine Command (TRADOC) usually tasks the TRADOC Analysis Center (TRAC) to conduct AoAs for ACAT I, ACAT IA, and ACAT II programs (U.S. Army, 1999, pp. 14 and 40), the CBTDEV is responsible for conducting the remaining ACAT II and III program AoAs if they are required by the MDA.[3]

The AoA is a document that reports the analytical evaluation of the performance, operational effectiveness, operational suitability, and estimated costs of alternative sys-

---

[3]   See the section "Role of a Combat Developer" in Appendix A.

From U.S. Army, 1999, p. 16:

The combat developer is that command, organizational element (including base operations and HQDA [Headquarters, Department of the Army]), and individual responsible for preparing and processing the materiel requirement document (MRD) and representing the user (organization and individual) of the new or modified system throughout the acquisition process.

tems to meet a mission capability or requirement that has been identified through the JCIDS process (Greenberg and Gates, 2006, pp. 1–2).[4] An AoA might show that all the alternative systems have unacceptable lifecycle costs, and it might require the Army to lower the capabilities required of the system in order to lower the lifecycle cost to an acceptable level. By introducing the consideration of lifecycle cost at this early stage (Milestone A), the Army can evaluate and monitor a system's performance-cost tradeoffs and make early design changes for affordability.

S&T Programs

DoD manages defense research through its six-stage research, development, test, and evaluation (RDT&E) program. The RDT&E program is divided into seven budget categories, which, with the exception of 6.6, RDT&E Management Support, correspond to RDT&E stages (see Table 2.1).

Budget categories and stages 6.1, 6.2, and 6.3 are referred to as the DoD S&T program. The S&T program's role is "to ensure our armed forces have superior and affordable technology to support their missions and to provide our troops with the tools they need to prevail in war" (Office of Science and Technology Policy, undated). S&T projects in the 6.1 Basic Research stage focus on producing new scientific or technological knowledge of interest to the military. Projects in the 6.2 Applied Research stage develop technologies for specific military applications. The 6.1 and even 6.2 projects

Table 2.1
RDT&E Budget Categories

No.	Category
6.1	Basic Research
6.2	Applied Research
6.3	Advanced Technology Development
6.4	Advanced Component Development and Prototypes
6.5	System Development and Demonstration
6.6	RDT&E Management Support
6.7	Operational Systems Development

---

[4] This AoA by Milestone A is the initial AoA. Greenberg and Gates (2006, pp. 1–2) point out that,

> in later phases of the acquisition process, the initial AoA is either updated or a new one is conducted, if warranted by then-existing circumstances. . . . An AoA might produce unacceptably high lifecycle costs for all alternatives . . . . Consequently, it might be necessary to lower requirements in order to reduce lifecycle costs to an acceptable level.

are in too early stages of technology development to be constrained by lifecycle cost consideration, as one wants to cast the net wide to capture technologies that have any possibility for military applications. However, even at these early stages, one should not preclude the consideration of a technology that aims to reduce weapon system cost. Projects in the 6.3 Advanced Technology Development stage support large-scale hardware and software development and demonstrate operational capability. As discussed throughout this report, lifecycle cost should be a key consideration at this stage.

The six RDT&E stages map to the first three phases in the Defense Acquisition Management Framework shown in Figure 2.1.[5] The S&T program corresponds to Pre–Systems Acquisition in Figure 2.1, which includes two sub-phases: CR and TD. Categories 6.4 and 6.5 correspond to the System Development and Demonstration phase in Figure 2.1.[6] One key purpose of pointing out the mapping between the DoD's new acquisition policy and the Army RDT&E process is to show that an AoA in preparation for an acquisition program's Milestone A should ideally occur during Advanced Technology Development (6.3) if not earlier during the Applied Research (6.2) program phase in the Army's S&T program.

As shown in Figure 2.1, initiation of DoD, including Army, acquisition programs currently occurs at Milestone B, the same as prior to the new acquisition policy introduced in May 2003 (U.S. Army, 2003, p. 29). In contrast, the AoA is now due at Milestone A, rather than Milestone B, as was the practice before 2003. Inertia and the time required for adjustment have resulted in some new programs being unable to perform the AoA earlier during the initial years after the new acquisition policy introduction in 2003. However, the Army wants to comply with the new DoD acquisition policy as fully as possible.

It is clear from the new acquisition policy that DoD wants AoA performed and lifecycle cost considered at Milestone A when the system is moved from CR to TD so that the technology and basic system design can be adjusted for affordability. For many acquisition programs, the CR and TD phases actually coincide with the S&T programs, particularly Applied Research (6.2) and Advanced Technology Development (6.3). To meet DoD's intent, we believe that lifecycle cost has to be considered during category 6.3, if not 6.2.

Unfortunately, the Army currently does not factor in lifecycle cost considerations during the S&T phase of development. While it has been said that S&T projects are not directed at developing specific operational weapon systems (see, for example, Moteff, 2003, p. 4), many current defense systems did evolve directly from an S&T

---

[5] Operational Systems Development (6.7) also maps to the acquisition process as part of the Production and Deployment phase and the Operations and Support phase in Figure 2.1.

[6] See, for example, U.S. Army, 1998, Figures I-11 and I-12. Moreover, while either the S&T community or the acquisition community could carry out category 6.4 work, the acquisition side traditionally prevails. See U.S. Government Accountability Office, 2006, p. 7.

project. For example, Portal Shield, an airbase and port biological detection system, came out of its Advanced Concept Technology Demonstration (ACTD) program, with the basic design and concept of operations intact. ACTDs are S&T projects whose aim is to exploit mature or maturing technologies to solve important military problems. Similarly, many current Army S&T projects should be expected to mature into fielded systems. Some examples are as follows:[7]

- Small-Unit Operations ATO, leading to a mobile networking wireless communications system (voice, data, video, and graphics)
- Overwatch ACTD, leading to a sensor targeting system that can provide real-time detection and classification of small-caliber and large-caliber munitions
- Unmanned Ground Mobility ATO, leading to near-autonomous unmanned systems
- Advanced Rotorcraft Technologies ATO, leading to an advanced rotorcraft system that can transition between rotary-wing and fixed-wing flight
- Future Tactical Truck System ACTD, leading to a tactical wheeled vehicle that is fuel efficient, transportable worldwide, and highly mobile; provides power and water generation, and survivability to the crew; and maintains low operation and support costs
- Joint Precision Airdrop System ACTD, leading to a high-altitude precision airdrop system
- Warfighting Physiological Status Monitoring ATO, leading to a wearable, modular yet integrated suite of sensors for remote monitoring of physiological parameters (heart rate, respiration, body temperature, body motion, and posture)
- High-Altitude Airship ATO, leading to a high-altitude airship.

We recognize that some Army S&T projects, even if they are already in the 6.3 stage, are aimed at providing technologies that are applicable to a wide range of weapon systems. For more general purposes, these projects can be treated in two ways. First, for some of these technologies, we can still measure their contributions to FOCs and implementation costs in the same manner as we measure those that are being developed for a specific system.[8] Thus, these technologies can be evaluated in the same manner as the system-specific technologies. Finally, even if some technologies are too

---

[7] These examples are from the Defense Technology Objective (DTO) data sheets shown in a Director of Defense Research and Engineering (DDR&E) report (DDR&E, 2006). All these examples are included as some of the 29 ATOs (also DTOs) used in this study. See Appendix B for more details about these examples and other ATOs used in the study.

[8] For example, the technologies developed under ATOs 6, 8, 16, 21, and 23 have applications to many systems. We estimate both their contributions to FOCs and remaining lifecycle costs in Appendixes B and C.

broadly scoped to be measurable by the methods suggested in this study,[9] our model will still provide the Army with an effective tool for managing the bulk of its S&T portfolio. Alternatively, these general ATOs can be analyzed separately by setting aside a portion of the S&T budget for them.

In sum, considering lifecycle costs as a part of S&T management should be important to the Army. We estimate that 11 of the 29 ATOs studied can develop major systems that will meet the RDT&E criteria (more than $365 million) and/or the procurement cost criteria (more than $2.19 billion)[10] to be in the ACAT I programs, which require AoA at Milestone A. An AoA will include lifecycle cost consideration. Moreover, AoAs are recommended for all other ACAT programs. Even if the Army can circumvent the requirement of considering lifecycle cost during the S&T phase, we believe that doing so is unwise, provided that there is a capability to incorporate such considerations. As stated in Chapter One, making smart design choices during S&T can prevent much of the need for costly or impractical redesign late in a development program. Further, even if S&T projects do not develop into major acquisition programs, the smaller projects together nonetheless have major financial implications for the Army. Considering lifecycle costs of these systems during their TD stage could also result in significant long-term savings. Finally, a uniform method for considering the lifecycle costs for most systems during their CR, TD, and S&T stages will allow the Army to make better decisions across its portfolio of development programs, making it better able to meet the materiel requirements of its soldiers and leaders at an affordable cost.

---

[9] We did exclude one ATO from our analysis for precisely this reason of non-system-specific technologies. See the exclusion in the section "ATOs Selected for Our Analysis" in Chapter Four.

[10] See Appendix A.

CHAPTER THREE

Description of Our Evolutionary Method

The genesis of the method and model in this study is the RAND PortMan R&D portfolio analysis and management method. PortMan had its beginnings in a decision framework developed for the U.S. Department of Energy's Industrial Materials for the Future program. For that study, the decision framework used an expected value (EV) based on the multiplication of three factors: (1) quantitative estimates of anticipated energy reduction benefits, (2) the potential of industrial materials to achieve required performance levels, and (3) the probability of the R&D project to be successful in the first place. EV is a measure of the attractiveness of alternative R&D projects (Silberglitt and Sherry, 2002). The framework provided a logical and auditable method for comparison of R&D projects within a portfolio based on best available data, as well as a means of reevaluating programs at a later date as R&D progresses.

The portfolio analysis method evolved from this decision framework and was used to analyze a portfolio of Applied Research (6.2) projects in a case study demonstration for the Office of Naval Research (Silberglitt et al., 2004). In that study, RAND brought together a panel of experts to estimate three factors: (1) the mean value to the military of the capability sought through R&D; (2) the extent to which the performance potential of the system to be developed, in the evaluation of the expert panel members, matched the level required to achieve the capability; and (3) the probability of success in transitioning the R&D project into a fielded system. The EV of each R&D project was then computed as the product of these three factors, and the R&D portfolio was evaluated in terms of the EVs, their components, and the uncertainties determined from the range of expert opinions.

Figure 3.1 is reproduced from another RAND monograph (Silberglitt et al., 2004). It shows the results of the PortMan case study demonstration for the Office of Naval Research. It is a plot, for each R&D project, of the product of the mean values, as determined by the expert panel, of capability and performance potential (y-axis) versus the mean values of the transition probability (x-axis). The EV for each project, which is the product of its x- and y-axis values, is shown in parentheses next to its data point, and the contours of equal EV are shown in the figure as a percentage of the maximum EV.

Figure 3.1
Portfolio of Navy Applied Research Projects from a Previous RAND Monograph

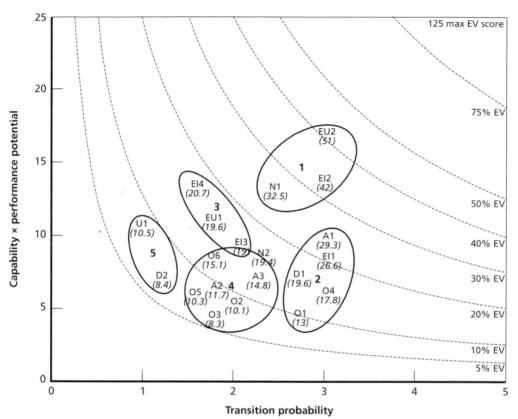

SOURCE: Silberglitt et al., 2004.
NOTES: The numbers 1 to 5 in the ellipses represent numbers assigned to groups of R&D projects that occupy similar positions on this plot. The maximum possible EV is 125, the product of the maximum value of 5 for each of the three factors. This corresponds to the maximum value of capability, with performance potential being 100 percent and transition probability also 100 percent. The EV contours, or isobars, are expressed as percentages of maximum EV. For example, the "30% EV" isobar represents a value that is 30 percent of 125 or 37.5 along the contour.
RAND MG761-3.1

This portfolio plot allowed the grouping of the R&D projects, as shown by the five ellipses in the figure. Projects in each group are similar in capability-performance potential and transition probability. Group 1 consists of the highest EV projects, which should be given the highest priority. Group 2 consists of relatively mature projects, as judged by the high transition probability, but that do not provide highly valued capabilities. These projects, representing incremental improvements, are candidates for shared funding with the offices for which they will provide systems developments in the near term. Group 3, on the other hand, represents projects that, if successful, could provide highly valued capabilities but that are at a relatively low level of maturity, as judged by the relatively low transition probability. The portfolio manager's job here is to try to improve the capability and/or transition probability of the project so that it moves into group 1, rather than group 2. This may mean, for example, providing resources and guidance to enable technical hurdles to be overcome so that performance levels can be met. It might also mean facilitating the development and execution of transition plans. Projects in this (high-value, high-risk) group could in fact be the key to filling capability gaps. Groups 4 and 5 can be similarly analyzed, and several projects here are candidates for decreased funding or termination under conditions of budget constraints.

While the previous versions of PortMan, as illustrated by the Navy example shown above, allowed detailed comparisons of R&D projects and provided guidance for the portfolio manager that was relevant to how to manage the R&D portfolio to maximize EV, investment strategy decisions were based on EV alone, with cost considerations based on the EV analysis introduced ex post facto. As described in the previous chapter, the objective of the current study is to incorporate cost analysis and constraints on an equal basis with the EV estimates and analysis, in order to minimize the total remaining lifecycle cost of the selected ATO projects and their derived systems[1] while at the same time meeting all FOC requirements.

Our Model

We use the current version of PortMan and a linear programming model to select a portfolio that consists of a subset of ATOs satisfying all the constraints and yielding the lowest total remaining lifecycle cost[2] for all the systems from the selected ATOs. Our model includes two classes of constraints. First, all FOCs must be fully met on an individual basis. The FOCs are actually FOC gaps, including improvements, which existing systems and their additional procured units cannot meet. These gaps are to be fully filled by new systems from ATOs. However, since traditionally the word "gap" is understood and frequently omitted, we follow this practice of using "FOC" and "FOC gap" interchangeably for ease of comparison with existing literature.[3]

Second, the total remaining S&T budget for the selected ATOs must not exceed a given budgeted amount that the Army can afford or is willing to pay. As a key application of this model, the budgeted amount will be varied to study the relation-

---

[1]   An ATO-derived system is defined as a weapon system or product that is developed under an ATO.

[2]   Only the future lifecycle cost is used, because past cost is already sunk and should not be used to inform any future decisions.

[3]   However, if possible confusion exists as to what we mean, we will state FOCs or FOC gaps explicitly in a footnote.

ship between the "affordable" total remaining S&T budget and the "affordable" total remaining lifecycle cost of the ATOs and their deployed systems. We will show that, in some cases, the Army may want to increase the total remaining S&T budget in order to significantly lower the total remaining lifecycle cost. Because the total remaining S&T budget is only a small portion of the total remaining lifecycle cost, some S&T projects including ATOs, although they have high S&T cost, can yield systems that are significantly cheaper to develop, demonstrate, acquire, operate, and maintain, thus lowering the total remaining lifecycle cost far more than the increase in the S&T budget required to fund these projects. In other words, since remaining S&T cost is already part of the remaining lifecycle cost, paying a higher S&T cost up front for those S&T projects can lead to a net savings in total remaining lifecycle cost.

The following three sub-sections describe the three steps of our method as applied to Army ATOs. Since this method is based on an extension of RAND's PortMan method, it uses an estimate of the EV of each ATO. However, this EV is defined in terms of contributions to individual FOCs, and these contributions are combined only when they address the same FOC but not when they address different FOCs. In contrast, previous applications of PortMan and the current typical method used by the Army S&T community combine the contributions for different FOCs in determining the attractiveness of a project, even though these contributions address different FOC requirements.

Step One: Attributing Value

Step one is to attribute the contribution of an ATO to various FOCs. Our procedure is to first define some terms and then show how they are related.

Definition 1. The contribution of ATO i to FOC j is the expected value, $E[V_{i,j}]$.

Definition 2. The required EV for FOC j is REV_j. This is the value that the Army wants all the funded ATOs to eventually contribute to FOC j.

Then, meeting an FOC requirement can be expressed by the following equation:

$$\sum_{i=1}^{n} x_i E\left[V_{i,j}\right] \geq REV_j \text{ for } j = 1, 2, \ldots, 13,$$

where

   x_i = 0 or 1, for a project that is not included or included in the selected portfolio, respectively, and

   n = number of ATOs (29 in our study).

This relationship allows the user of our method to define an REV for *each* FOC, and thus select a project portfolio that simultaneously meets all of these required values. The user can also change the requirements to see how the change affects the portfolio selection.

To determine EV scores for ATOs, we rely on the FOC gaps, which we assume to be the same as the sub-requirements listed in the 11 FOC descriptions in an authoritative and standard document (U.S. Army, 2005b). However, we made two modifications to the sub-requirements. First, we divided several of the sub-requirements to make them more mutually exclusive so that they do not overlap and are not duplicative of each other. Second, in cases for which the same sub-requirement appeared in more than one FOC, we merged them, again to eliminate duplication. In such cases, the score for this sub-requirement will be placed in only one FOC. Finally, we use 13, instead of the original 11, FOCs because we broke the FOC, *protection*, into three FOCs, *personnel protection, asset protection,* and *information protection,* to emphasize the importance of protection in counterinsurgency and the technological differences in executing three different types of protection.

Definition 3. The contribution of ATO i to sub-requirement k of FOC j is the expected sub-requirement value, $E[V_{i,k,j}]$.

The EV score for the ith ATO on the jth FOC, or simply the $EV_{i,j}$ score, is then defined by:

$$EV_{i,j} = E\left[V_{i,j}\right] = \sum_{k=1}^{m} E\left[V_{i,k,j}\right],$$

where there are m sub-requirements in FOC j.

A useful parameter is the total EV score from all ATOs on the jth FOC or simply the EV_j score:

$$EV_j = E\left[V_j\right] = \sum_{i=1}^{n} E\left[V_{i,j}\right].$$

The FOCs and their sub-requirements used in our model are from a TRADOC pamphlet (U.S. Army, 2005b) and are listed as follows:

FOC 1—Battle Command

1. Layered, integrated command and control (C2) for joint, multinational, and interagency operations on the move
2. Integrated tactical network with universal worldwide access to the Global Information Grid (GIG)
3. Networked force optimized for mobile operations
4. Decision planning and support capabilities
5. Information operations integrated with information management and ISR (intelligence, surveillance, and reconnaissance)
6. Information protection and rapid restoration of information and information systems.

FOC 2—Battlespace Awareness

1. C2 of battlespace awareness assets
2. Ability to observe and collect information worldwide
3. Analysis of intelligence information
4. Ability to model, simulate, and forecast
5. Ability to manage knowledge
6. Fusion of information.

FOC 3—Mounted-Dismounted Maneuver

1. Effective battle command
2. Unsurpassed battlespace awareness
3. Exceptional air-maneuver support
4. Dependable and accurate line of sight/beyond line of sight/non–line of sight (LOS/BLOS/NLOS) lethality
5. Ability to deploy rapidly for strategic response
   - Sustainment with minimal load and logistics footprint
   - Quality, realistic, accessible training
   - Human engineering for improved solider-system interface
   - Enhanced mobility and ability to shoot in multiple directions simultaneously
   - Operations in urban and complex terrain.

FOC 4—Air Maneuver

1. Responsible and sustainable aviation support
2. Effective aviation operations in contemporary operating environments
3. Reconnaissance, surveillance, and target acquisition and attack operations
4. Assured and timely connectivity with the supported force
5. Mounted vertical maneuver.

FOC 5—LOS/BLOS/NLOS Lethality

1. LOS/BLOS lethality via precision, networked, responsive fires of extended range at any time, in any place, while minimizing fratricide and non-combatant casualties
2. NLOS lethality seamlessly from tactical to operational distances with no gaps or loss of timeliness, with advanced automated fire control and distribution to sort out high payoff and most-dangerous targets rapidly and in depth.

FOC 6—Maneuver Support

1. Provide assured mobility
2. Deny enemy freedom of action
3. Engage and control populations, including enemy prisoners of war and other detainees
4. Employ non-lethal effects
5. Detect and neutralize environmental hazards
6. Reduce environmental damage from military operations
7. Understand the battlespace environment.

FOC 7—Personnel Protection

1. On-body personnel protection
2. Off-body personnel protection.

FOC 8—Asset Protection[4]

1. On-asset asset protection
2. Off-asset asset protection.

FOC 9—Information Protection

1. On-system information protection
2. Off-system information protection.

FOC 10—Strategic Responsiveness and Deployability

1. Airlift and sealift assets and enablers
2. Theater access enablers, a responsive distribution system, and installations as U.S. flagships.

FOC 11—Maneuver Sustainment

1. Improved sustainability
2. Global precision delivery enhancements
3. Power and energy
4. Enhancements in readiness, reliability, maintainability, and commonality for sustained operational tempo
5. Global force health and fitness (disease prevention is not counted here, but rather in FOC 7)

---

[4]   An asset can be a weapon system, a vehicle, a building, etc.

6. Global casualty care management/evacuation
7. Global casualty prevention (disease prevention is not counted here, but rather in FOC 7)
8. Improved soldier support
9. Global military religious support.

FOC 12—Training, Leadership, and Education

1. Leadership training and education
2. Accessible training
3. Realistic training
4. Responsive training development
5. Training for joint, interagency, intergovernmental, and multinational (JIIM) tasks
6. Trainability
7. Managing unit performance
8. Providing universal training support.

FOC 13—Human Engineering

1. Reduce soldier dismounted movement approach load to no more than 40 pounds and fighting load to 15 pounds
2. Decrease task complexity and execution times, while minimizing sensory, cognitive, and physical demands on each soldier
3. Provide mobility enhancements through environmental ride quality and task automation
4. Exploit unmanned technology in manned systems to enhance continuous 24-7 operations.

Gap Space Coverage. To estimate the contribution of ATOs to FOCs, we require a metric that represents the degree to which specific FOC requirement gaps are filled by specific ATOs. We approach this problem by first estimating the extent to which the capabilities supplied by each ATO, if it successfully meets its objectives, will address each FOC requirement gap. We then define the *gap space* as the area spanned or covered by all gaps, and the ATO's *gap space coverage* as the sum of all gaps addressed by that ATO. We can then use the gap space coverage of an FOC by an ATO as an estimate of the degree to which that ATO addresses that FOC.

We use the following approach to estimate the FOC gap space coverage of ATOs. First, we recognize that FOCs apply to warfighters in three different situations: (1) off the battlefield; (2) on the way to the battlefield; and (3) on the battlefield, as indicated in Figure 3.2. Accordingly, we assign a factor of one-third to the ATO's coverage of each of these situations. Next, we divide the FOC gap space into a mutually exclusive

Figure 3.2
Situations in Which FOCs Apply to Warfighters

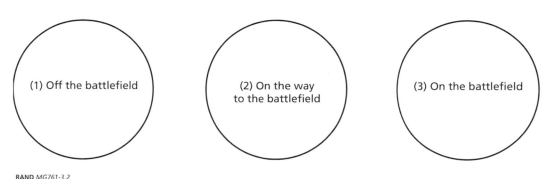

(1) Off the battlefield

(2) On the way
to the battlefield

(3) On the battlefield

RAND *MG761-3.2*

set of categories tailored to each FOC, as indicated in Table 3.1. We then estimate the ATO's coverage of the FOC's gap space as the fraction of that FOC's categories addressed by that ATO. Finally, since we do not have an explicit description of the gaps, we assume that each FOC sub-requirement represents a gap, and we multiply the ATO's gap space coverage estimate by the fraction of that FOC's sub-requirements that the ATO addresses.

It is important to note that since an explicit description of gaps was not available, it was also not possible to estimate the precise extent to which an ATO addresses a gap, even though we can count the number of sub-requirements addressed. As a result, we introduce the concept of coverage score ($CV_{i,j}$), which indicates the extent to which the ATO addresses the FOC gaps, but not the quality of the coverage for any given ATO. Our estimate of the coverage score of the ith ATO on the jth FOC is given by

$$CV_{i,j} = C[V_{i,j}] = S_i \times C_j \times SR_j,$$

where S_i is the fraction of situations shown in Figure 3.2 that ATO_i addresses, C_j is the fraction of categories shown for FOC_j in Table 3.1 that ATO_i addresses, SR_j is the fraction of sub-requirements defined (in a TRADOC pamphlet [U.S. Army, 2005b]) for FOC_j that ATO_i addresses.

To convert coverage score into EV, we simply assume that when any ATO satisfies an FOC gap, it does so at a level of 50 percent.[5] Thus,

---

[5] Implicit in this estimate is the assumption that a successful ATO makes a meaningful contribution to filling the FOC gaps that it addresses, because it has achieved its defined payoffs according to its defined metrics. For this demonstration of our method, we assume that these "meaningful" contributions fill 50 percent of all gaps that they address. This 50 percent is an assumed value for the scaling factor used in previous PortMan versions. This factor is called the *technical potential,* which is a measure of the degree to which the payoffs are actually achieved based on the performance level actually achievable by the ATO. For details, see Silberglitt and Sherry, 2002; and Silberglitt et al., 2004. Within the current Assistant Secretary of the Army (Acquisition, Logistics and Technology) portfolio analysis approach, an expert panel estimates the technical feasibility of each ATO. This estimate of technical feasibility could be used as a surrogate estimate of the PortMan *technical potential.* The tech-

Table 3.1
Categories of FOC Gap Space

FOC	Basis for Categories	Categories
1. Battle command	Must provide	Command, control, communications, and computers
2. Battlespace awareness	Must provide common operating picture	LOS fire, B/NLOS fire, enemy force location, and hazard location
3. Mounted-dismounted maneuver	Needed for maneuver	Forces, mobility, weapons, and equipment and supplies
4. Air maneuver	Needed for maneuver	Forces, mobility, weapons, and equipment and supplies
5. LOS/BLOS/NLOS lethality	TRADOC defined	LOS, BLOS, and NLOS
6. Maneuver support	Needed for maneuver	Forces, mobility, weapons, and equipment and supplies
7. Personnel protection	Must protect against	Disease[a] and injury[b]
8. Asset protection	Must protect against	Accident, LOS fire, B/NLOS fire, and hazards
9. Information protection	Must protect against	Theft, destruction, modification, and access restriction
10. Strategic responsiveness and deployability	Must provide	Readiness, transportation, and delivery
11. Maneuver sustainment	Needed for maneuver	Forces, mobility, weapons, and equipment and supplies
12. Training, leadership, and education	Must train or educate concerning	Doctrine, equipment, people, and environment
13. Human engineering	Must engineer[c]	Systems, people or tasks, and human-system interfaces

[a] Disease is further divided into malaria, HIV/AIDS, tuberculosis, emerging infectious diseases (EIDs, e.g., dengue fever), and influenza.

[b] Injury is further divided into accidents, LOS fire, B/NLOS fire, and hazards (e.g., mines, booby traps, and improvised explosive devices [IEDs]).

[c] If the engineering is to reduce soldier load (e.g., see U.S. Army, 2005b, paragraph 4-73), then the categories are (a) food, (b) clothing and shelter, (c) weapons, and (d) equipment and supplies.

$$\mathrm{E}\left[V_{i,j}\right] = \tfrac{1}{2}\mathrm{C}\left[V_{i,j}\right].$$

The data that we used to estimate the ATO CVs and EVs were derived from the Defense Technology Objectives, 2006 (DDR&E, 2006) since all of the ATOs we

nical potential varies between 0 and 100 percent. For demonstration of the methodology in this study, we use the same 50 percent for every ATO.

chose for this example application of our method and model were also DTOs. Appendix B describes the rationale for our CV estimates, based on published DTO data, the equation above, and the definitions presented in Figure 3.2 and Table 3.1. A numerical example is also provided in the section "Degree of Meeting FOC Requirement" in Chapter Four.[6]

Step Two: Marginal Cost

Step two of the procedure is to estimate the remaining lifecycle cost over that of the legacy system (marginal remaining lifecycle cost [MRLCC]).[7] It is a marginal cost that is to be incurred in the future, if the ATO is funded and the system based on it is fielded, over the cost of the legacy system. In other words, we start by asking to what extent the REV for *each* FOC can be met by legacy systems alone[8] and what would be the total remaining lifecycle cost for these systems to serve over a planning period.[9] Once this baseline is established, we use the FOC gaps determined by the Army[10] and make sure that we will use only the parts of the gaps that will not be met by legacy systems. Then, two aspects of each of the existing ATOs are measured: (1) each ATO's contribution to filling the individual FOC gaps and (2) the remaining lifecycle cost of completing the S&T project (ATO in question) and fielding the system(s) derived from the ATO in question.

If the system will produce a capability that does not exist in the legacy system, we can attribute to this new system all of the contribution to gap filling, as well as 100 percent of the remaining lifecycle cost. At the other extreme, if the system does not provide any new capability, but can lower the remaining lifecycle cost of the corresponding legacy system that provides the same capability, we will assign zero to the new system's capability-gap contribution, and a negative cost to its MRLCC. This

---

[6] However, it should be emphasized that this study focused on methodology development. To demonstrate the methodology, we took a retrospective look at how well a small subset of the 2005 ATO projects could meet a hypothetical set of gaps, which we assumed occurred in every FOC requirement and sub-requirement. In other words, the study did not use data on real capability gaps and provided no information on how well the Army S&T portfolio meets the actual gaps. Moreover, since the ATOs were evaluated against hypothetical gaps, one should not draw any conclusions about the merits or drawbacks of any specific S&T project discussed in this study in meeting real Army capability gaps.

[7] The term *marginal remaining lifecycle cost* is used so often in this report that, for brevity, the word "marginal" is often omitted because *remaining lifecycle cost* is always estimated and understood, in this study, in a marginal way.

[8] Legacy systems are existing systems, additional purchases of these systems, and planned systems that have not used the technologies developed under the existing ATOs.

[9] For this study, we assume that the new systems derived from the ATOs or additional legacy systems will be acquired over a planning period of 20 years. Once a system is fielded, it will be operated and maintained until it is retired, and the lifecycle cost includes all costs until it is retired.

[10] FOC gaps are determined by the Army Capabilities Integration Center, which conducts an annual capabilities-based FOC needs analysis coordinated with the Army Headquarters and the Joint Staff.

negative cost is equal to the cost savings if this system were used instead of its corresponding legacy system.

It is important for any model or method to capture cost savings of a new system, since this would be an important consideration when selecting an ATO. Also, the cost savings should be properly captured in the cost category of remaining lifecycle cost, as opposed to being treated as a capability gap contribution and mixed with other contributions to FOCs ($E[V_{i,j}]$), because cost savings and capability gap contributions are not even measured in the same units. The general case occurs when the new system derived from the ATO in question can be installed on a legacy system, and the combined system will help meet a fraction of the FOC gap that all the legacy systems together still cannot meet. In this case, the remaining lifecycle cost of the new system over that of the legacy system is the marginal cost to the legacy system to address the FOC gap.[11]

To determine the marginal remaining lifecycle cost of each ATO, we first define the following terms:

- Marginal remaining lifecycle cost (MRLCC) is the cost difference between the new system's remaining lifecycle cost and that of the legacy system.
- Remaining S&T cost (RSTC) is the future cost to complete the ATO program.
- System Development and Demonstration cost (SDDC) is the cost to develop and demonstrate the new system derived from the ATO in question.
- Marginal procurement cost (MPC) is the cost of acquiring new systems, minus the cost of buying legacy systems instead, to serve the planning period.
- Marginal upgrade cost (MUGC) is the cost to modify the new system for maintaining currency over that of the legacy system.
- Marginal operating and maintenance cost (MOMC) is the cost difference between servicing the new systems over the planning period versus servicing the legacy systems. For some systems, a major contributor to MOMC is the marginal manpower cost (MMC), which is the manpower cost in operating the new system during its operating life over that of the legacy system. The manpower operating cost is equal to the annual manpower cost times the operating life. The annual manpower cost is personnel time spent in operating the system during the year times the salary rate.
- Discount rate (DR)[12] is the interest rate to discount future costs and to make them comparable to the current costs.

---

[11] The marginal cost is the cost differential between the new system and the corresponding legacy system. While this cost is often positive (i.e., added cost), it can be negative if the new system produces a net cost savings to the system. In that case, the total new system cost will be lower than that of the corresponding legacy system.

[12] We assume 0 percent discounting here for simplicity, because this study is a demonstration, not an actual application, of the method. When one applies our method to a real case, one should consider using a discount rate such as the 3 percent rate typically used by DoD.

- Marginal unit cost (MUC) is the unit cost of a new system minus that of the legacy system to be replaced.

Then,

$$MRLCC = RSTC + SDDC + MPC + MUGC + MOMC.$$

Moreover,

$$MPC = UCNEW \times NPNEW - UCLEG \times NPLEG,$$

where

UCLEG is the unit cost of the legacy system,
UCNEW is the unit cost of new system,
NPLEG is the number of additional legacy systems that need to be procured to provide service for the planning period,
NPNEW is the number of units of new systems procured during the planning period.
For a common case in which the number of new units is the same as the number of legacy systems being replaced,

$$MPC = MUC \times NPNEW.$$

Also,

$$MOMC = UCNEW \times NPNEW \times OMCNEW\% - UCLEG \times NPLEG \times OMCLEG\%,$$

where OMCNEW% is the operating and maintenance cost during the operating life of the new systems expressed as a percentage of the procurement cost for the new systems, and OMCLEG% is the operating and maintenance cost during the operating life of the legacy systems expressed as a percentage of the procurement cost for the legacy systems.

Step Three: Linear Programming Model

Step three of the procedure is to use a linear programming model to yield the lowest total remaining lifecycle cost to meet all FOC requirements on an individual basis. The linear programming model selects a package of ATO projects so that the cost to complete these selected ATO programs and to develop, field, and operate their resulting systems to meet all FOC requirements is minimized. This can be expressed as follows:

$$\text{Minimize } \sum_{i=1}^{n} x_i \text{ MRLCC}_i$$

$$\text{subject to } \sum_{i=1}^{n} x_i \text{ RSTC}_i \leq \text{BSTC}$$

$$\text{and } \sum_{i=1}^{n} x_i \text{E}\left[V_{i,j}\right] \geq \text{REV}_j \text{ for } j \text{ from 1 to 13 (FOCs)},$$

where

    x_i is 0 or 1 for non-selected and selected ATOs, respectively, with i the ATO program number running from 1 to n, where n is 29 in this study,

    RSTC_i is the remaining S&T cost for the ATO program i,

    BSTC is the budgeted total remaining S&T cost for all the selected ATO projects,

    MRLCC_i is the remaining lifecycle cost, which includes RSTC_i, for ATO program i,

    REV is the required expected value.

Simplifying Assumptions Used

In this study, we have made two simplifying assumptions. First, we assume that all ATOs are at a level of maturity such that the probability of the system(s) being successfully derived from each of these ATOs is 100 percent. Second, we use a single-point estimate for all input parameters, such as an ATO's contribution (EV) to each FOC requirement, a system's unit cost, the number of units fielded, and the operating and maintenance cost. The follow-on study that we began in October 2007 has expanded the model in this study to allow for the probability of success *not* being 100 percent.[13]

Potential of the Model

We have developed our model with the potential for much broader applicability under various situations different from those demonstrated in this monograph. First, we have designed the model to capture the key drivers, both technical and economic, for making informed decisions about selection of new projects, or selection of existing projects for continued support at any milestone, for a single service, or among services' projects within DoD. The model is applicable to inform decisions at any milestone, because the model calculations can be based on the future or remaining lifecycle cost with respect to any desired decision point.

---

[13] The probability of success can be based on the *technology readiness level* of the ATO project. In the follow-on study, we use a simulation model to examine how well the random combinations of project successes will meet the required value for each FOC. The same model can be used to study multiple-point or range estimates for other input parameters.

Second, the model user can employ rough estimates of input parameters to obtain a first-order or general picture of the alternative decision choices. Then, as more accurate input data become available, the user can improve the accuracy of the model predictions.

Third, we use a linear programming model, because of its convenience in accommodating additional constraints that may be called for in certain applications. For example, in an AoA, there are generally several alternative projects, Y_1, Y_2, . . . Y_p, for meeting a certain capability gap. An AoA is typically aimed at selecting only one for development.[14] A linear programming model can limit the selection to at most one project by adding an internal constraint of $Y_1 + Y_2 + \ldots + Y_p \leq 1$, where Y_i takes the value of 1 if the model recommends the project to be selected for development, and 0 if it is not to be selected. Moreover, this constraint also applies to a case where all Y_1, Y_2, . . . Y_p are incompatible with each other so that only one project should be selected. While this feature is not currently available, it can be incorporated into the model should such a situation arise.

However, the selection in an AoA is further complicated by two factors. These alternative projects may also contribute to gaps other than the gap in question. The selection should take account of these benefits. The other factor is that projects undertaken primarily for other gaps may contribute to the gap in question as well. Again, the selection needs to account for these benefits.

---

[14] The linear programming model can also be generalized to include the consideration of the costs and benefits of selecting more than one project for development.

CHAPTER FOUR

Estimation of Input Parameters

We emphasize again that this study has two key objectives: presenting a mathematical framework and suggesting ways to estimate the input parameters, including cost components. The Army S&T community very likely has better supporting data than was available to us and that will allow better estimates of input parameters. Thus, in the event that the reader does not want to use our cost estimation approaches, we suggest that he or she consider the adaptation of our framework independently; we believe that the framework is an appropriate way to quantify the contribution and cost of projects and to perform an effective project portfolio analysis.

In this chapter, we describe how various input parameters are estimated. These parameters, which have been defined in Chapter Three, are

- coverage score (CV_{ij}), which represents the percentage of gap space to which an ATO contributes
- expected value (EV_{ij}), which is the CV_{ij} times the quality of the contribution[1]
- remaining S&T cost (RSTC)
- SDDC
- unit cost of a new system (UCNEW)
- number of units of new systems to be acquired over the planning period (NPNEW)
- marginal upgrade cost of the systems over the planning period (MUGC)
- marginal operating and maintenance cost of the systems fielded (MOMC).

ATOs Selected for Our Analysis

Instead of discussing how we estimate the input parameters to our model abstractly, we use a real, but simplified, case. The Army S&T Master Plan described 172 ATOs (U.S.

---

[1] This study assumes the quality of a contribution to be 50 percent. See the subsection "Gap Space Coverage" in Chapter Three.

Army, 2005c).[2] They are the fundamental building blocks of the Army S&T programs and the highest-priority efforts in the 6.2 Applied Research and 6.3 Advanced Technology Development programs. There are 70 ATOs in the 6.2 program and 102 in the 6.3 program. Seven of the 6.3 ATOs are also named as advanced technology demonstrations (ATDs). ATDs illustrate the potential for enhanced military operational capability or cost-effectiveness, help TRADOC develop more informed requirements, and assist the materiel developer to assess and reduce risk before committing to full-scale System Development and Demonstration (U.S. Army, 2005c, p. B-1). Another nine of the 6.3 ATOs are ACTDs. These demonstrations are designed to expedite the transition of maturing technologies from the developers to the users. ACTDs emphasize technology assessment and integration, as opposed to TD. The goal is to provide a prototype capability to warfighters and allow them to evaluate the capabilities in real military exercises and at a scale that is sufficient to fully assess military utility (U.S. Army, 2005c).

Since this study is one of methodological development, only a subset of existing ATOs are included to demonstrate the method. Because the ATOs in the advanced technology program are better developed and the lifecycle costs for the systems to be eventually developed from them are better known, we selected ATOs from this program for a demonstration of the modified PortMan method and its applications. Further, we selected all 29 such ATOs that are also DTOs.[3] These ATOs, being also the technology objectives for the military as a whole, are even more crucial for meeting the key FOC gaps.

The current method used by the Army Deputy Assistant Secretary for Research and Technology to evaluate ATOs focuses on their contributions to FOCs. Under that method, the attractiveness of an ATO depends on the sum of its contributions to various FOCs. While this approach places emphasis on the overall contributions of ATOs to meeting FOC capability gaps, it leads to the danger that certain FOC capability gaps will be unmet. In other words, it provides no basis for assurance that the portfolio of ATOs will meet the individual requirements of each FOC.[4] With respect to S&T funds, the approach wants to ensure that the funds are spread more or less evenly to

---

[2]   This was the latest master plan available when this study commenced.

[3]   Actually, there are 30 ATOs that are also DTOs. However, the one called "Army/DARPA Enabling Technologies for Future Combat Systems" consists of 13 projects, which are to develop critical technology improvements for Future Combat Systems (FCS) platform variants and the network. Further, FCS has four components—manned ground vehicles, unmanned systems, FCS network, and soldiers—and includes a large number of platforms and systems. We consider that this ATO applies to many Army systems, as opposed to one or a class of specific systems. It was not included in our analysis because of the difficulty of estimating its remaining lifecycle cost for the many different new and improved systems over that of their legacy systems. See the section "S&T Programs" in Chapter Two for a discussion of situations in which an ATO is more appropriately treated separately from the rest of the ATOs in a portfolio analysis.

[4]   See Chapter One for additional discussion of this point.

address all 13 FOCs. Unfortunately, the method does not identify the optimal distribution of S&T funds, because meeting some FOC capability gaps requires more S&T funding while others can be addressed with relatively fewer funds. Moreover, since lifecycle cost is not currently incorporated into the portfolio analysis of the ATOs, the Army cannot presently assess the attractiveness of ATOs in terms of how they contribute to reducing the total remaining lifecycle cost. The modified PortMan model that is the subject of this monograph can help the relevant Army offices address these issues and, as such, will provide a valuable and complementary capability to the Army's S&T planning tool set.

Overwatch ACTD

We use the Overwatch ACTD (ATO#3 in the Army S&T Master Plan) to illustrate how estimates are made for our model input parameters for each of the 29 ATOs. We also describe additional estimation methods for various components of the lifecycle cost. Since different ATOs may have different availability of data on which an estimate may be based, having several estimation methods enhances the chance that one can make an estimate based on whatever data are available. The example used here, with further details provided in Appendixes B, C, and D, should suffice for illustrative purposes to demonstrate how input parameters for each of the 29 ATOs were estimated.[5]

Objective

The objective of the Overwatch ACTD is to develop a device that can detect and locate hostile, LOS fire originating out to a range of 600 meters, with an accuracy of 3 meters for small arms at maximum range and within 2 seconds of firing. This device will be mounted on both high-mobility multipurpose wheeled vehicles (HMMWVs) and unmanned ground vehicles.

Degree of Meeting FOC Requirement

To estimate how well Overwatch meets FOC requirements, we used the following data sources:

- The July 2005 Army S&T Master Plan, which was the latest version available during the time of our study.

---

[5] However, it should be emphasized that this study focused on methodology development. To demonstrate the methodology, we took a retrospective look at how well a small subset of the 2005 ATO projects could meet a hypothetical set of gaps, which we assumed occurred in every FOC requirement and sub-requirement. In other words, the study did not use data on real capability gaps and provided no information on how well the Army S&T portfolio meets the actual gaps. Moreover, since the ATOs were evaluated against hypothetical gaps, one should not draw any conclusions about the merits or drawbacks of any specific S&T project discussed in this study in meeting real Army capability gaps.

- A report on the 2006 Defense Technology Objectives (DDR&E, 2006). This report was the latest available during the time of our study. DTOs are updated regularly on this DDR&E Web site. We selected ATOs that are also DTOs for this study, because more data are available on DTOs.

As described in Chapter Three, we use a slightly modified version of the FOCs described in a TRADOC pamphlet (U.S. Army, 2005b). Based on the data on objectives, metrics, payoffs, challenges, and milestones in the ATO data sources listed above and on the definitions of the FOCs, we concluded that the Overwatch ACTD contributes to meeting capability requirements in seven of 13 FOCs:

- FOC 2, battlespace awareness
- FOC 3, mounted/dismounted maneuver
- FOC 6, maneuver support
- FOC 7, personnel protection
- FOC 8, asset protection
- FOC 11, maneuver sustainment
- FOC 13, human engineering.

Since the Overwatch device will be mounted on HMMWVs and unmanned ground vehicles, according to Figure 3.2 it applies to situations (2) and (3), on the way to the battlefield and on the battlefield. Therefore, we multiply its gap coverage estimates by two-thirds. The following describes our estimates of the coverage score of Overwatch with respect to each of the FOCs listed above.

FOC 2: Battlespace Awareness. Of the four categories in Table 3.1, Overwatch applies only to LOS fire, because its sensors are all LOS sensors. Therefore, our gap coverage estimate is one-fourth. According to TRADOC pamphlet 525-66 (U.S. Army, 2005b), FOC 2 has six sub-requirements, and Overwatch contributes to only one of these, "C2 of battlespace awareness assets." Therefore, we multiply our gap coverage estimate by one-sixth. Thus, our estimate of the CV contribution of Overwatch (ATO#3) to FOC 2, $C[V_{3,2}]$, is

$$C[V_{3,2}] = 2/3 \times 1/4 \times 1/6 = 0.0278.$$

FOC 3: Mounted-Dismounted Maneuver. Of the four categories in Table 3.1, Overwatch applies only to two categories ("forces" and "weapons"), so our gap coverage estimate is one-half (i.e., 2/4 or 1/2). According to TRADOC pamphlet 525-66 (U.S. Army, 2005b), FOC 3 has ten sub-requirements. Overwatch contributes to three of these, "unsurpassed battlespace awareness," "dependable and accurate LOS/BLOS/NLOS lethality," and "operations in urban and complex terrain." However, as noted in the previous paragraph, Overwatch is an LOS device, so we apply an extra factor of one-fourth to "unsurpassed battlespace awareness" (i.e., 1/10 × 1/4) and an extra factor of one-third to "dependable and accurate LOS/BLOS/NLOS lethality" (i.e., $1/10 \times 1/3$). In addition, Overwatch contributes fully to "operations in urban and complex terrain" (i.e., $1/10 \times 1$). Accordingly, our estimate of $C[V_{3,3}]$ is

$$C[V_{3,3}] = 2/3 \times 1/2 \times 1/10 \times (1/4 + 1/3 + 1) = 0.0528.$$

FOC 6: Maneuver Support. Of the four categories in Table 3.1, Overwatch applies only to "forces" and "weapons," so our gap coverage estimate is one-half. According to TRADOC pamphlet 525-66 (U.S. Army, 2005b), FOC 6 has seven sub-requirements. Overwatch contributes to two of these, "assure mobility," and "deny enemy freedom of action." However, as noted in the previous paragraph, Overwatch is an LOS device, so we apply an extra factor of one-fourth to both sub-requirements. Accordingly, our estimate of $C[V_{3,6}]$ is

$$C[V_{3,6}] = 2/3 \times 1/2 \times 2/7 \times 1/4 = 0.0238.$$

FOC 7: Personnel Protection. Of the two categories in Table 3.1, Overwatch applies only to "injury," so our gap coverage estimate is one-half. Moreover, of the four categories for "injury" defined in the notes of Table 3.1, Overwatch applies only to "LOS fire," so we multiply by an additional factor of one-fourth. There are no sub-requirements for this FOC, but we introduce an additional factor of one-half to provide redundancy for protection of the warfighter. Accordingly, our estimate of $C[V_{3,7}]$ is

$$C[V_{3,7}] = 2/3 \times 1/2 \times 1/4 \times 1/2 = 0.0417.$$

FOC 8: Asset Protection. Of the four categories in Table 3.1, Overwatch applies only to "LOS fire," so our gap coverage estimate is one-fourth. There are no sub-requirements for this FOC, but we introduce an additional factor of one-half to provide redundancy for protection of warfighting assets. Accordingly, our estimate of $C[V_{3,8}]$ is

$$C[V_{3,8}] = 2/3 \times 1/4 \times 1/2 = 0.0833.$$

FOC 11: Maneuver Sustainment. Of the four categories in Table 3.1, Overwatch applies only to two categories ("forces" and "weapons"), so our gap coverage estimate is one-half. According to TRADOC pamphlet 525-66 (U.S. Army, 2005b), FOC 11 has nine sub-requirements. Overwatch contributes only to "improved sustainability." However, as noted previously, Overwatch is an LOS device, so we apply an extra factor of one-fourth. Accordingly, our estimate of $C[V_{3,11}]$ is

$$C[V_{3,11}] = 2/3 \times 1/2 \times 1/9 \times 1/4 = 0.00926.$$

FOC 13: Human Engineering. Of the three categories in Table 3.1, Overwatch applies only to engineering of the "system," so our gap coverage estimate is one-third. According to TRADOC pamphlet 525-66 (U.S. Army, 2005b), FOC 13 has four sub-requirements. Overwatch contributes only to "decrease task complexity." Accordingly, our estimate of $C[V_{3,13}]$ is

$$C[V_{3,13}] = 2/3 \times 1/3 \times 1/4 = 0.0556.$$

Table 4.1 lists the EV contributions[6] for all 29 ATOs used to demonstrate our method. Because an FOC can receive contributions from multiple ATOs, the EVs for an FOC can be summed to exceed 100 percent. Appendix B provides information similar to that presented above for Overwatch concerning the coverage scores for the other 28 ATOs.

Estimation of Cost Components

We describe below how we estimated the cost components of the device derived from the Overwatch ACTD.

Remaining S&T Cost. The remaining S&T program cost is $1.6 million, which is stated on the corresponding DTO data sheet in a DDR&E report (DDR&E, 2006)[7] and is the latest data on remaining S&T program cost available on the DDR&E Web site.

System Development and Demonstration Cost. The annual Army RDT&E Budget Item Justification Sheets for fiscal year (FY) 1999 to FY 2009 (see, e.g., U.S. Army, 2008) prepared by the Assistant Secretary of the Army (Financial Management and Comptroller) show the system development and demonstration (SDD) projects and their costs (SDDCs). These costs correspond to actual programs planned or already under way. However, only the more successful ATOs are selected for SDD. For our study, we need to address the "what if" issue before we can select the most cost-effective ATOs for continued funding. That is, What if an ATO continued into the SDD phase, what would be the SDDC? For those ATOs that are not funded for the SDD phase, the aforementioned data source does not provide the hypothetical cost that we need for our study.

In principle, one can use an approach that is based on analogy to estimate the SDDC. The RDT&E costs of existing systems that are judged to be similar to the system in question can be used to obtain the SDDCs. One good source of these data is the annual *U.S. Weapon Systems Costs* by Data Search Associates.

---

[6]   Note that the table shows EVs, which are half of coverage scores. See the subsection "Gap Space Coverage" in Chapter Three.

[7]   The data sheet listed both S&T funding from the individual services and from DoD. For this study, only funding from the Army was included, under the assumption that other funding was used to meet the capability requirements of the other services.

Table 4.1
Estimations of Contributions of ATOs to FOCs (in percentage)

ATO#	EV 1	EV 2	EV 3	EV 4	EV 5	EV 6	EV 7	EV 8	EV 9	EV 10	EV 11	EV 12	EV 13
1	2.78	1.39	1.67			1.19							0.93
2							2.50						
3		1.39	2.64			1.19	2.09	4.17			0.46		2.78
4						3.57	4.17	8.35			5.55		
5	5.55	4.17	5.00				8.35	16.65			2.78		11.10
6			3.34								3.71		
7	8.35	5.55	1.67										1.39
8			3.34								7.40		
9						3.57					8.35		
10				.						22.20	7.40		
11											5.55		
12											1.39		
13							2.50				4.17		
14							2.50				2.78		
15							9.40				5.55		
16							3.13				5.55		
17		8.35	5.00			3.57							
18							2.50				2.78		
19										16.65	11.10		
20	8.35		1.25								1.39	31.25	
21	2.78	1.39	4.17		2.78	1.19	1.04				1.85	1.04	1.39
22	4.17										1.39	9.40	
23	4.17	8.35	7.50	5.00			3.13	6.25					2.78
24		16.65				3.57	9.40	18.75					
25	. 2.09			2.50					25.00				
26	33.35	16.65	7.50		16.65	5.35	6.25	12.50			4.17		
27	2.09	6.25									1.39		
28	4.17		1.25	2.50							2.78		
29			1.67			4.76	2.09	4.17			3.71		

NOTES: Numbers in this table, as well as elsewhere in this monograph, are shown with more places than their significant figures to allow for others to replicate and check our calculations. Blank cells in the table mean that there is no contribution.

For our study, since many ATO systems do not have similar existing systems, we use an approach that is based on the development and use of a universal multiplier. For example, the Overwatch device can be considered to be an electronics system. For each of 34 existing electronics systems in the *U.S. Weapon Systems Costs*, we expressed the Test and Evaluation (T&E), or SDD, cost as a multiplier of the R&D budget, or 6.1 to 6.3 program cost.[8] Averaging these multiples yielded a universal multiplier of 1.44. The details in developing this multiplier, as well as other universal multipliers and curves, are described in Appendix D. The SDDC is the total S&T cost times the multiplier.

In the case of Overwatch, while the remaining S&T cost is $1.6 million, the total S&T cost or the cost from start to finish of Overwatch is $8.9 million.[9] We used this alternative approach to obtain $13 million as the SDDC (i.e., $8.9 million × 1.44).

Marginal Upgrade Cost. Even after a weapon system is fielded, modifications to correct minor design defects and upgrades to improve performance will continue, and the costs of this work should be estimated and included in the PortMan model. Because this cost is not included in the DoD or Army budget projections for ATOs, we estimate it using a multiplier approach similar to that for estimating SDDC above, except it is a multiplier to the acquisition cost.[10]

Overwatch is a new system, not a replacement of any legacy system. Thus, the marginal upgrade cost (over that of the replaced legacy system) is the same as its upgrade cost. We estimated this cost to be $2.5 billion.

Unit Cost. Unit costs can be estimated by analogy to similar systems found in the *U.S. Weapon Systems Costs*. Adjustments to these costs to account for technical and other design differences are applied to get a final estimate of the unit cost for the system in question.

In some cases, the unit cost is explicit in the *Procurement Justification Books* published by the Assistant Secretary of the Army (Financial Management and Comptroller).[11]

In the case of Overwatch, we estimated the unit cost to be $1.6 million, which is half as much as the existing Advanced Targeting Forward Looking Infrared (ATFLIR) sensor system (Nicholas and Rossi, 2006, pp. 3–12) found in the *U.S. Weapon Systems Costs, 2006*. (See Appendix C for details.)

Number of Units. There are two approaches for estimating the number of systems that will be fielded. The first is to examine how the system will be used and by whom. Based on these, an estimate of the number of systems required can be determined.

---

[8] Other sister publications to *U.S. Weapon Systems Costs* provide costs for additional systems.

[9] The Overwatch project started in FY 2004 and the total S&T (or R&D) cost until and including FY 2007 is $8.9 million.

[10] The multiplier is determined in the section "Upgrade Cost Projections" in Appendix D.

[11] The latest being the FY 1999 to FY 2009 edition.

The second approach is by analogy. We first identify existing systems that are performing missions similar to those to be served by the new system in question, and then estimate the number of new systems that would be required to serve a planning period of 20 years. Since the acquisition of additional units for an existing system often is still ongoing, an extrapolation is required to project the total number of units acquired over a 20-year period. Based on 34 electronics systems still being procured in 2006, we obtained an annual purchase curve, with the number of annual purchases expressed as a multiple of the average annual purchases during the first three years of introduction (see Figure 4.1). These 34 electronics systems approximately follow a universal curve independent of specific systems. With annual purchases expressed in this way, one can extrapolate from the number of units already acquired for an existing system to the total number of existing systems required to serve the 20-year planning period.

The following example illustrates the calculation. Let us assume a perfect case in which the new electronics system to be derived from an ATO will serve exactly the same missions (but better) as an existing electronics system that has been fielded for eight years. The number and acquisition schedule of new systems required to serve the planning period of 20 years are assumed to be the same as those of the existing systems that the new systems would replace. Further assume that the annual purchases for the

Figure 4.1
Annual Purchases of an Electronics System over a 20-Year Period

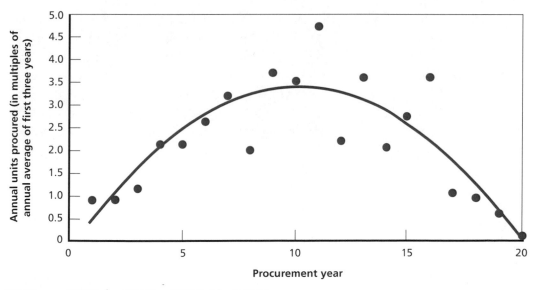

NOTES: $y = -0.0346x^2 + 0.7097x - 0.2804$. $R^2 = 0.7092$.

This figure represents an introductory rate of a new support equipment system (based on existing systems still being acquired in 2006).

RAND MG761-4.1

existing system are 28, 50, and 114 during the first three years of introduction, and the cumulative purchases thus far (i.e., during the first eight years) are 1,500. Thus, we know that 1,500 new units would be required during the first eight years. The remaining question is how many more new units would be acquired over the remaining 12 years of the planning period. Because the existing system has an acquisition history of only eight years, we need data from similar electronics systems to estimate the acquisition plan for the remaining 12 years. Our estimation of the number of new systems to be acquired over a planning period of 20 years starts with determining the average annual purchases during the first three years to be 64 (i.e., average of 28, 50, and 114). Next, we determine from the universal curve for electronics systems in Figure 4.1 that the cumulative purchases from year nine to year 20 are 1.67 times those during the first eight years (1,500 units). Using the same multiplier, we project the purchases of the existing systems from year nine to year 20 to be 1.67 times 1,500, or to be 2,498, and the total purchases for the 20 years to be 3,998 units. More details can be found in Appendix C.

The advantages of this analogy approach are twofold. First, the number of units can be easily determined from publications such as the annual *U.S. Weapon Systems Costs* or the *Procurement Justification Books*. Second, the actual procurement stream of units for an existing system over time has automatically reflected the need for spares, replacement of retired units, and the timing of obsolescence. A disadvantage is that there is not always an existing system similar enough to the developmental one to make a reasonable comparison.

However, for the Overwatch example, we were able to examine how the system will be used and by whom, so that we used the first approach described above to estimate the number of Overwatch devices needed. This device for detecting and locating hostile fire will likely be mounted on ground vehicles. About 17,000 HMMWVs are being deployed in Iraq. In a convoy operation, there is no need to have the device on every vehicle. We assume that one in ten will be so equipped. Further, all devices first deployed will have one replacement during the 20-year deployment period. We also assume 20 percent more devices will be purchased for use on other platforms and in other theaters. Rounding the estimate to the nearest thousand, we obtained 4,000 as the number of units to be procured.

Marginal Operating and Maintenance Cost. As described previously, the determination of MOMC involves an estimate of OMCNEW%, which is the operating and maintenance (O&M) cost during the operating life of the new systems expressed as a percentage of the acquisition cost for the new systems.[12] We also need to estimate OMCLEG%, which is the operating and maintenance cost of the legacy system that the new system is going to replace.

---

[12] See the section "Step Two: Marginal Cost" in Chapter Three.

As described in Chapter Three, the O&M cost during a system's operating lifetime is expressed as a percentage of the unit acquisition cost. In theory, this estimate can be based on the O&M costs of existing systems similar to the system in question. These O&M costs can be derived from the data on consumables and repairables appearing in the Operating and Support Management Information System (OSMIS).[13] In practice, it is difficult to aggregate the O&M costs for a specific existing system, because the costs are often recorded for different combinations of the system's components, instead of the specific system in question.

As a demonstration of our method, we started with two O&M costs that OSMIS was able to provide on a system base. We found the costs of consumables and repairables for a Single Channel Ground to Air Radio System (SINCGARS) during the unit's operating life to be 150 percent of the unit cost. We used this as our "highest" O&M cost estimate. We also found the costs of consumables and repairables for an Advanced Field Artillery Tactical Data System (AFATDS) to be merely 10 percent of the unit cost. We used this as our "lowest" O&M cost estimate. Between these two extremes, we inserted three intermediate O&M costs at equal increments: 115 percent denoted "higher" cost, 80 percent denoted "reference" cost, and 45 percent denoted "lower" cost. Then, for any system under consideration, we assigned the O&M cost to one of these five levels based on the characteristics of the system and its operating environment.

Because the Overwatch device involves sensitive sensors, which need to be kept clean and calibrated, and yet will have to operate in an outdoor environment, we assigned to it the "higher" O&M cost, 115 percent of unit cost. Moreover, there is no legacy system for Overwatch to replace, and thus there is no OMCLEG%. Therefore, the marginal upgrade cost (over the legacy system) is the same as the upgrade cost, which is the unit cost ($1.56 million) times the number of units procured (4,000) times the O&M multiplier (1.15), which equals $7.2 billion.

Manpower Operating Cost. The manpower cost for operating the system during its operating life is equal to the annual manpower cost times the operating life. The annual manpower cost is personnel time spent operating the system during the year times the salary rate.[14]

---

[13] The following text is quoted from the OSMIS Web site (U.S. Army, Cost and Economic Analysis Center, undated):

> The Operating and Support Management Information System (OSMIS) is the core of the Army Visibility and Management of Operating and Support Costs (VAMOSC) program. OSMIS tracks operating and support information for over one thousand major Army weapon/materiel systems for the Office of the Deputy Assistant Secretary of the Army for Cost and Economics (DASA-CE). OSMIS-tracked systems include combat vehicles, tactical vehicles, artillery systems, aircraft, electronics systems, and miscellaneous engineering systems. DASA-CE generates and fields a wide variety of requests for OSMIS data to support analyses tasks throughout the defense community.

[14] Note that the manpower for maintaining and repairing the system is included in the O&M cost.

Table 4.2
Estimations of Cost Components for 29 ATOs

A	B	C	D	E	F	G	H	I
ATO Project No.	Marginal Unit Cost (MUC)	No. of New Systems Procured (NPNEW)	Remaining S&T Cost (RSTC)	System Development & Demonstration Cost (SDDC)	Marginal Procurement Cost (MPC)	Marginal Upgrade Cost (MUGC)	Marginal O&M Cost (MOMC)	Marginal Remaining Lifecycle Cost (MRLCC)
1	$0	265,607	$17.8	$40	$0	$0	–$319	–$261
2	$0.000005	1,716,000	$17.2	$70	$9	$3	$1	$100
3	$1.56	4,000	$1.6	$13	$6,252	$2,541	$7,190	$15,997
4	$0.1	4,000	$52.0	$269	$400	$163	$320	$1,204
5	$0.09	6,109	$55.5	$381	$556	$226	$567	$1,786
6	$1	900	$25.0	$167	$900	$366	$2,369	$3,827
7	$14	29	$45.2	$372	$406	$165	$325	$1,313
8	$2.81	16,800	$95.9	$138	$47,212	$19,186	–$9,442	$57,189
9	$0.02	182,483	$2.0	$30	$4,133	$1,680	–$2,755	$3,090
10	$0.15	1,000	$0.0	$0	$150	$61	$173	$383
11	$0.0005	186,890	$2.8	$50	$93	$38	$98	$283
12	$0.000004	133,426	$2.9	$162	$1	$0	$0	$165
13	$0		$7.7	$45	$0	$0	$0	$53
14	$0.000005	1,716,000	$8.5	$54	$9	$3	$1	$75
15	$0		$5.3	$13	$0	$0	$0	$18
16	$0		$3.6	$22	$0	$0	$0	$25
17	$55	50	$0.0	$0	$2,750	$1,118	$2,888	$6,755

Table 4.2—Continued

A ATO Project No.	B Marginal Unit Cost (MUC)	C No. of New Systems Procured (NPNEW)	D Remaining S&T Cost (RSTC)	E System Development & Demonstration Cost (SDDC)	F Marginal Procurement Cost (MPC)	G Marginal Upgrade Cost (MUGC)	H Marginal O&M Cost (MOMC)	I Marginal Remaining Lifecycle Cost (MRLCC)
18	$0.000005	2,058,528	$59.2	$295	$10	$4	$1	$370
19	$5.99	1	$7.7	$46	$6	$2	$7	$69
20	–$0.0022	1,000	$0.0	$0	–$2	–$1	$0	–$3
21	$0		$79.9	$370	$0	$0	$0	$450
22	$0		$15.1	$46	$0	$0	$0	$62
23	$0		$32.1	$83	$0	$0	$0	$115
24	$3	608	$0.0	$0	$1,824	$741	$2,098	$4,663
25	$55	1	$7.4	$77	$55	$22	$0	$162
26	$0		$48.3	$111	$0	$0	$0	$159
27	$0		$22.1	$51	$0	$0	$0	$73
28	$0.008	6,667	$10.3	$16	$53	$22	$88	$189
29	$0		$20.4	$49	$0	$0	$0	$70

NOTES: Letters in parentheses in the equation below refer to corresponding columns in the table. All columns are in millions of dollars except columns A and C, which are dimensionless.

MRLCC (I) = RSTC (D) + SDDC (E) + MPC (F) + MUGC (G) + MOMC (H)

There are two extremes in terms of manpower requirements. The low extreme is occupied by a partially or fully automatic system mounted on a platform such as a HMMWV. Since the soldiers are dedicating their time to many other tasks, the time spent on an automatic system can be negligible. We assume the manpower cost in this extreme case to be zero. The high extreme is occupied by a system that is manually operated and carried by soldiers, who are totally dedicated to the handling of the system. If these soldiers are spending a fraction of their working hours dedicated to the performance of this system, we will assign the same fraction of their annual salary to the annual manpower cost for the operation of this system. In all cases, we assess only the marginal cost, namely, the difference in manpower cost between the system in question and the legacy system being replaced. In the case of a new system for a new capability, there is no legacy system and no legacy cost. Then, the marginal cost is equal to the cost of the new system.

Because the Overwatch system will be installed on ground vehicles and its detection and location of hostile fire are automatic, we consider its manpower cost to be close to the low extreme and assume it to be zero.

Remaining Lifecycle Costs for 29 ATOs

As explained above, the best way to estimate a component's cost depends on the type of system derived from the ATO in question. In Appendix C, we have used the various methods described above to estimate these costs for each of the 29 ATOs selected for demonstrating the method and its applications. Our cost estimation for all 29 ATOs is summarized in Table 4.2. The remaining S&T costs are from a DDR&E report (DDR&E, 2006),[15] while all of the other cost components are estimated by us.

---

[15] As explained in the subsection "Estimation of Cost Components" in this chapter, we simply looked up the future or remaining annual S&T costs from the DDR&E report.

CHAPTER FIVE

Applications to S&T Portfolio Management

In this chapter, we describe how the method and model developed in this study can be applied to address various S&T issues concerning project selection, portfolio monitoring, and the optimality of the S&T budget.

Depth in Capability Contributions in the ATO Pool

The first question for the Army to ask is whether supporting all existing ATOs to their completion would meet all 13 individual FOC requirements. If these requirements are not met, more ATOs would have to be developed. On the other hand, if existing ATOs are meeting requirements, how much extra capability above the FOC requirements are these ATOs providing? In later sections of this chapter, we show how much remaining S&T budget or remaining lifecycle cost the Army could save by reducing these extra capabilities. If the savings are not large, the Army might want to keep the extra capabilities as an additional safety margin. Moreover, if the Army is forced to cut funds, the model will show the different amounts of savings that can be obtained by reducing different FOC requirements. Knowing these different impacts, the Army will be in a better position to decide where to make the cuts among individual FOC requirements.

Figure 5.1 shows the fractions of individual FOC requirements met, defined in our methodology as the EV_j of each FOC_j, if all of the 29 ATOs we selected for our method and model application demonstration are funded. Since these ATOs are only a small subset of the 172 existing ATOs, one should not expect this subset to meet 100 percent of each of the 13 FOC requirements.[1] Yet, considering Figure 5.1, these 29 ATOs together are on average meeting 46 percent of the FOC requirements.[2] Sev-

---

[1] The Army has 11 FOC requirements, but we split one of them into three to total 13. See the section "Step One: Attributing Value" in Chapter Three.

[2] If one adds up the 13 fractions for the 13 FOCs in Figure 5.1 and divides the sum by 13, one gets 0.46 or 46 percent. Mathematically, this average is

$$\Sigma_{j=1}^{13} E\left[V_j\right] \div 13.$$

Figure 5.1
Degrees of FOC Requirements Met by 29 ATOs

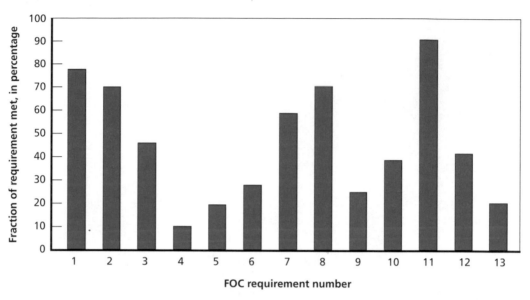

enteen percent would be a fair share of the requirements for the 29 ATOs to meet—29 ATOs are only 17 percent of the existing 172 ATOs (i.e., 29 ÷ 172). In other words, our 29 selected ATOs need to contribute, on average, only 37 percent of their potential (46 percent) in order to meet their fair share of 17 percent (i.e., 37 percent times 46 percent equals 17 percent). For the purpose of demonstrating applications of our model, we assume that the individual REV_j's for these 29 ATOs are 40 percent (a rounding up from 37 percent) of the actual EV_j's shown in Figure 5.1. From here on, the 40 percent requirement is also referred to as the reference requirement when we study the impacts of changing requirements.

Based on the above discussion, we have cast the 29 ATOs, although only a portion of the 172 existing ATOs, into a form that can be used to simulate the complete existing ATO set. This example is now restated as follows for our demonstration. There are 29 existing ATOs. If all of them are funded and their systems are fielded, they will provide the individual FOC capabilities shown in the blue bars in Figure 5.2. In addition, the individual FOC requirements (REV_j) that they need to fill (under the assumptions described above) are shown as the red bars in Figure 5.2.[3] The difference between the blue and red bars indicates the depth of the ATO pool. The larger the dif-

---

[3] To repeat, for the selected set of 29 ATOs in Figure 5.2, each red bar or REV for a given FOC is 40 percent (rounded up from 37 percent, as discussed above) of the actual expected values, $E[V_j]$ for the corresponding FOC. Further, these red bars averaged over 13 FOCs are 17 percent of 100 percent of the FOC requirements for the full set of 172 ATOs.

Figure 5.2
Achievable EV and REV for 29 ATOs

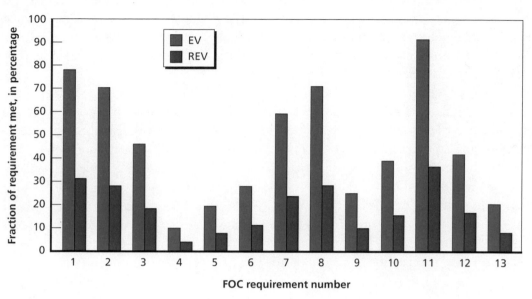

RAND *MG761-5.2*

ference, the more power the ATO pool has in meeting and exceeding the capability requirements.

Inadequacy in Existing ATO Pool

If some blue bars were below the corresponding red bars in Figure 5.2, the Army would need to either cut back on FOC requirements or add more ATOs.[4] The graph will indicate which individual FOCs that the existing ATOs fail to fully meet and thus which existing ATOs should be expanded or in which areas new ATOs need to be added. Further, the model can be run with the expanded and new ATOs to see whether all the individual requirements are then met. Thus, the model can also be used to size the extent to which the ATO pool needs to be expanded.

---

[4] While inadequacy cannot happen in our demonstration with 29 ATOs, since the individual requirements are always less than the potential contributions (i.e., 40 percent), one could encounter such an inadequacy when examining the full set of existing ATOs.

Optimal ATO Set

Using the 40-percent requirement level discussed previously, we ran the model to find the subset of these 29 ATOs that at least meet all individual FOC requirements (shown as red bars in Figure 5.2 and Figure 5.3) at the lowest total remaining lifecycle cost. This subset, consisting of ATOs 1, 5, 13, 15, 16, 19, 20, 23, 25, 26, and 29, is the optimal ATO set for the 40-percent requirement level. As shown in Figure 5.3, these 11 ATOs not only meet the REV_j's, but exceed these REV_j's by various amounts.[5] The portfolio manager would want to treat these 11 ATOs as more important than the others, since the fielding of systems based on these ATOs will yield the lowest total remaining lifecycle cost at which all FOC capability gap requirements are met.[6]

In this study, we have assumed that all ATOs, if funded to completion, will be successful in the sense that the systems derived from them will produce the EVs for their contributions to FOCs and will have remaining lifecycle costs as expected. In this case of sure success, only the selected 11 ATOs need to be funded, and one can terminate the other 18 ATOs. In reality and in the ongoing follow-on study, each ATO has a non-zero probability of failure or not meeting the expectation. In that case, the

Figure 5.3
Achievable EV and REV for the Optimally Selected 11 ATOs

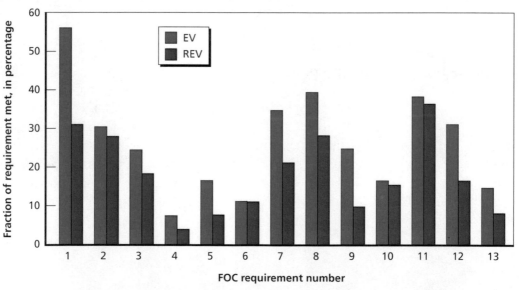

RAND MG761-5.3

---

[5]   In contrast, Figure 5.2 shows all 29 ATOs, as opposed to the optimally selected 11 ATOs in Figure 5.3.

[6]   Here we are referring to the requirements based on 40 percent of the actual expected values of the 29 ATOs chosen for a demonstration in this study.

other 18 ATOs would serve the purpose of backups in the event one or more selected ATOs fail. Considering the reality of the situation, one should view the 11 selected ATOs as higher priority than those not selected. In the event of a cut in total remaining S&T budget such that not all 29 ATOs can be supported, one should consider dropping ATOs in the unselected group and protecting those in the selected group. In the subsection below "ATO Ranking," we will refine this prioritization by considering uncertainties in future FOC requirements.

Monitoring of the Existing ATO Portfolio

The model can be run periodically, say annually, to alert the portfolio manager about the current status of the existing ATO projects. For example, are there any ATOs that are no longer needed, because some other ATOs have performed better than previously anticipated and are likely to fill the future capabilities better? On the other hand, do some deficiencies in meeting REV_j now appear, because some ATOs are not progressing as well as anticipated? Which ones are causing the deficiencies? Can the progress of these ATOs be brought back to normal, or is there a need to bring in new ATOs to make up for these deficiencies? In the event of a budget cut, which ATOs should be downsized or even eliminated?

Remaining S&T Cost Attributable to Each FOC

We now return to Figure 5.3 to study how one should attribute the remaining S&T cost of a particular ATO among 13 FOCs. In other words, we want to know what portion of an ATO's remaining S&T cost should be counted as the cost of the ATO's contribution to a particular FOC. Let us start with a simple hypothetical example. Suppose there are two ATOs that contribute to two FOCs. ATO One (ATO1) will need $3 million more to complete its S&T phase. When the systems derived from this ATO are eventually acquired and fielded, they are expected to contribute an EV of 75 percent to FOC 1 and an EV of 50 percent to FOC 2. On the other hand, ATO Two (ATO2) will need $5 million more to complete its S&T phase. Its systems will contribute an EV of 40 percent to FOC 1 and 80 percent to FOC 2. We have examined various schemes to allocate the remaining S&T costs to FOCs and adopted the following allocation as being logical and reasonable. ATO1 and ATO2 together have contributed an EV of 115 percent to FOC 1 (i.e., 75 percent + 40 percent), and 130 percent to FOC 2 (i.e., 50 percent + 80 percent). The 75 percent contribution from ATO1 accounts for 65 percent (i.e., 75 percent ÷ 115 percent) of the total contribution to FOC 1, and the 50 percent contribution from ATO1 accounts for 38 percent (i.e., 50 percent ÷ 130 percent) of the total contribution to FOC 2. Finally, we allocate the $3 million proportionally to

ATO1's relative contributions to FOC 1 and FOC 2. Thus, the allocation is $1.9 million to FOC 1 (i.e., $3 million × 65 percent ÷ (65 percent + 38 percent)) and $1.1 million to FOC 2 (i.e., $3 million × 38 percent ÷ (65 percent + 38 percent)). Similarly, the allocation for ATO2 is $1.8 million to FOC 1 and $3.2 million to FOC 2.

For this study, all ATOs are assumed to be successfully completed if they are funded to complete their S&T phase. Then, there is no need to continue funding the other 18 ATOs.[7] We now want to know how the remaining S&T cost of the 11 projects should be attributed to each of the 13 FOCs. We first apply the cost allocation method above to ATO#5, Warfighter-Systems Interaction, which is one of the 11 ATOs selected by the model to yield the least total remaining lifecycle cost. Its derived system will contribute to FOCs 1, 2, 3, 7, 8, 11, and 13, as shown in Table 5.1. Consider FOC 2 as an example. At the reference requirement, the total EV contribution to FOC 2 from all 11 ATOs is 30.56 percent. Therefore, the 4.17 percent contribution to FOC 2 from Warfighter-Systems Interaction is 13.63 percent (4.17 percent ÷ 30.56 percent) of total contributions to FOC 2 as shown in Table 5.1. This is referred to as the relative contribution of Warfighter-Systems Interaction to FOC 2. The total relative contribution of ATO#5 to all FOCs is 194.46 percent. Since the remaining S&T cost for ATO#5 (from its DTO data sheet—DDR&E, 2006) is $55.50 million, we attribute a cost of $3.89 million (13.63 percent × $55.50 million ÷ 194.46 percent) to FOC 2.

Table 5.1
Allocation of ATO#5's Remaining S&T Cost Among 13 FOCs

FOC_j	EV_j from ATO#5 in %	Universal	Allocation Fraction for ATO#5 in Decimals	Distribution of #5 Remaining S&T Cost, in $million	
1	5.55%	56.28%	9.86%	2.8145	4.17 ÷ 30.56
2	4.17%	30.56%	13.63%	3.8904	($55.50 × .1363) ÷ 1.9446
3	5.00%	24.58%	20.34%	5.8056	
4		7.50%			
5		16.65%			
6		14.43%			
7	8.35%	31.71%	26.33%	7.5153	
8	16.65%	39.57%	42.08%	12.0105	
9		25.00%			
10		16.65%			
11	2.78%	38.41%	7.24%	2.0659	
12		31.25%			
13	11.10%	14.81%	74.97%	21.3979	
ATO Total	53.60%	347.38%	1.9446	$55.50	

---

[7]   In the ongoing study, we examine the impact of ATOs that have a possibility of failure and the benefits and costs of keeping additional ATOs as backup.

Repeating the same allocation procedure for the other ten selected ATOs, we arrive at the allocation of the remaining S&T budget among the 13 FOC$_j$'s shown in Figure 5.4. The S&T funds attributable to individual FOCs are not distributed uniformly. To develop the capability for the 13th FOC requires the largest amount of S&T funds, while the 12th FOC requires the least.[8] On the other hand, the method used by the Army tends to distribute S&T funds evenly to cover all 13 FOCs. This study does not endorse this practice, because some FOCs need a relatively larger share of the S&T funds to develop the systems that provide such FOCs.

Cost-Effectiveness of S&T Funds in Developing Individual FOCs

In the above section, we have shown that it takes different amounts of remaining S&T funding[9] to develop or contribute to different individual FOCs. Moreover, since these FOCs are of different magnitudes, it should be of interest to know the remaining S&T funds needed to develop each unit (i.e., percentage or fraction in this study)

Figure 5.4
For Development, Some FOCs Need More S&T Funds

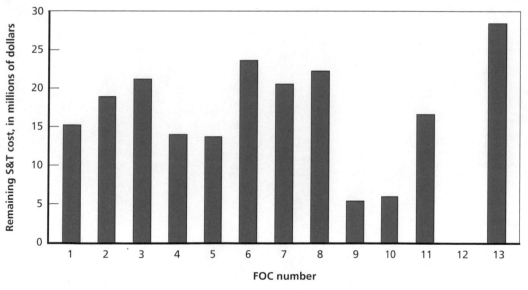

NOTE: Dollar amounts reflect the remaining S&T cost attributable to each FOC.
RAND *MG761-5.4*

[8] In fact the cost is zero, because only ATO#20 is needed to meet the requirement of FOC 12 and has zero remaining S&T cost.

[9] We do not include past S&T costs that have already been consumed.

of capability for the various FOCs. To determine this cost-effectiveness, one simply divides the remaining S&T cost in Figure 5.4 by the total EV_j score in Figure 5.3 for the corresponding FOC. The results are shown in Figure 5.5. The cost ranking based on remaining S&T cost (Figure 5.4) and the remaining S&T cost per unit of capability are not necessarily the same. The 12th FOC is the least expensive to develop in both cases (Figures 5.4 and 5.5).[10] However, the top five ranked FOCs are different. For the least amount of S&T cost to develop capability that at least meets the requirement, the top five are FOCs 12, 9, 10, 5, and 4, respectively (see Figure 5.4). For lowest S&T cost to develop each percentage point of capability, the top five are FOCs 12, 9, 1, 10, and 11, respectively (see Figure 5.5). Both cost rankings can be used to guide allocation of remaining S&T funds among ATOs. The former (Figure 5.4) indicates the level of S&T funding needed in order to develop capabilities for meeting the individual FOC requirements, as an alternative to allocating S&T funds evenly (and likely suboptimally) among FOCs. The latter (Figure 5.5) indicates that, if there is a cut in the remaining S&T budget, reducing which FOCs would yield the largest savings in S&T budget. For the same percentage cut in an FOC, the cut in FOC 6 would yield the largest savings in remaining S&T budget because this capability necessitates the highest remaining S&T budget to develop. It should be emphasized that these two cost

Figure 5.5
Some FOCs Need More S&T Funds to Develop Each Percentage Point of Capability

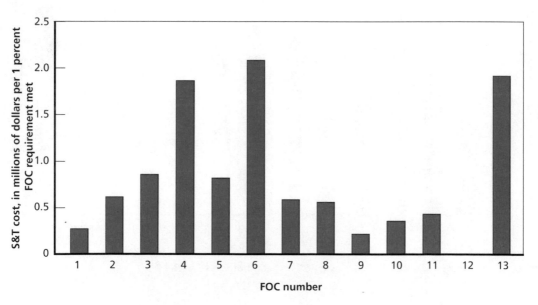

---

[10] Again, this is zero. At the reference requirement, this FOC is completely satisfied by ATO#20, which has no remaining S&T cost associated with it.

rankings should be used in conjunction with actual model runs with varying remaining S&T budget and FOC requirements, since each provides rough and complementary information, whereas the model output captures the complicated and combined effects to give much more accurate results.

ATO Ranking

Since the future Army budget for total remaining lifecycle cost and the future strategic environment are uncertain, the requirement may turn out to be higher or lower than the reference. For example, slower economic growth can lead to a lower budget. Further, a lower budget would force a cut in requirements. A new strategic environment, such as that present post–9/11, can lead to additional requirements. ATOs that are selected to yield the lowest total remaining lifecycle cost regardless of future requirement levels should be ranked highest and protected.

Between the 1988 and 2007 fiscal years, the annual defense budget averaged $437 billion and fluctuated between $345 billion and $564 billion per year. The lowest yearly budget is 21 percent below the average, while the highest yearly budget is 29 percent above it.[11] Because we use a weapon deployment period of 20 years, the uncertainty in budget and requirement averaging over a 20-year period is likely to be smaller than the annual uncertainty, and we assume the former to be 20 percent. The question can now be stated as follows:

> Given a reference requirement of 40 percent, when the FOC requirements are set at 0.8 (i.e., 32 percent), 0.9 (36 percent), 1 (40 percent), 1.1 (44 percent), and 1.2 (48 percent) of the reference requirement, which ATOs are selected most often in these five cases in order to yield the least total remaining lifecycle cost?[12]

Those ATOs selected most often are most important because no matter what the future will be, they are needed to yield the lowest total remaining lifecycle cost.

In Figure 5.6, a green cell indicates that an ATO was selected under the particular FOC requirement. A red cell indicates that the ATO was not selected. ATOs 1, 5, 15, 16, 19, 20, 23, 25, 26, and 29 are always selected under every requirement level (also see Table 5.2). They are the highest ranking ATOs and should be protected in the event of a budget cut.

ATO#10 is selected at two requirement levels, while ATOs 4, 13, 21, and 27 are selected once. These ATOs are still needed in some future possible requirement levels.

---

[11] These numbers have been calculated from data provided in U.S. Department of Defense, 2007, Table 6-1, pp. 62–67.

[12] The *reference requirement* refers to the 13 REVs (red bars) in Figure 5.2. At 0.8RR, we mean that all 13 individual requirements are reduced by 20 percent or set at 80 percent of the reference REVs.

Figure 5.6
ATOs That Could Meet Requirements at the Lowest Remaining Lifecycle Cost

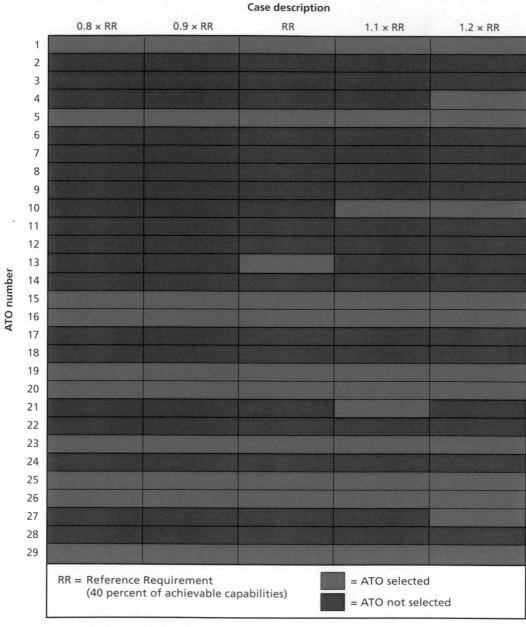

To meet the requirements at all levels at the lowest total remaining lifecycle cost, they, as the second-rank ATOs, should be protected as well. The remaining 14 ATOs are never selected, and they form the lowest rank. In this study, all ATOs are assumed to be successfully completed if they continue to be funded. This assumption also means

Table 5.2
Selection of ATOs Across Requirement Levels (within 20 percent of the reference requirement)

Selection Count	ATO Number (1 through 29)
5	1, 5, 15, 16, 19, 20, 23, 25, 26, 29
4	Not applicable
3	Not applicable
2	10
1	4, 13, 21, 27
0	2, 3, 6, 7, 8, 9, 11, 12, 14, 17, 18, 22, 24, 28

that there is no need for any backup ATOs, since the selected ATOs will not fail. In our ongoing study, all ATOs are assumed to have a possibility of failure, and we examine the role of these never-selected ATOs serving as backup in the event that the selected ones fail.

One may wonder why a particular ATO, such as 13, is selected at the reference requirement level, but not selected at other levels. This results from the complicated interplay of and competition among all ATOs' costs and contributions to the individual capability requirements. The linear programming model used automatically calculates the optimal selection in the midst of these complexities.

We conduct two sensitivity tests in Appendixes E and F. In Appendix E, we study a future in which the Army requirements are more uncertain. The model shows which additional ATOs should be funded to completion in order to be able to meet a wider range of future requirements. We also study whether rougher-grain cost data (grouping our costs in dollars into seven levels) would lead to different model results. In Appendix F, we found in this case that the project selection is identical to that shown in Figure 5.6. This result suggests that perhaps it will be useful to use a much quicker and easier Delphi method, such as that used in earlier PortMan work to estimate EVs, for cost estimation. If the Delphi method is proven to be applicable, our model may find wider adoption by various agencies in the S&T and acquisition community.

Cost Reduction Required for Moving to a Higher Rank

We demonstrated in the previous section the ranking of the ATOs. A higher-ranking project is a more essential project. When there is a cut in the total S&T budget, there should be a higher propensity to continue funding a higher-ranked ATO. Thus, it is of interest to know how the lifecycle cost of the system derived from an ATO needs to be reduced in order for the ATO to move up in rank.[13]

We use ATO#8, Power and Energy, as an example. The ATO seeks to develop hybrid electric technology for Future Combat System vehicles. Because of the large number of vehicles to be produced, as well as the cost of hybrid technology, the proposed weapon system carries the highest MRLCC estimate of any ATO at $57.2 billion. It is never selected under any capability requirement in Figure 5.6. However, assume that a breakthrough in hybrid technology occurs, and the cost over legacy of the system can be reduced from $2.81 million to $390 thousand, while maintaining the same savings in O&M over the legacy system. The MRLCC is reduced to only $6.35 million in that case, and ATO#8 is selected at the reference requirement level. Thus, the model can provide cost reduction targets to S&T program managers if they wish to move the ATO up in rank.

Recommended Rule for Allocating Lifecycle Funds to Develop FOCs

Similar to S&T funds allocation, there is a tendency to cover all FOCs by distributing evenly the remaining lifecycle funds, which are used to pay the remaining S&T cost; SDDC; and system acquisition, operation, and maintenance costs. This study does not endorse this practice of uniform funding, because some FOCs need a relatively larger share of the funds than others to develop their capabilities.

Our model will determine the optimal fund allocation. The model will find a subset of ATOs that will meet all individual requirements fully at the lowest total remaining lifecycle cost. As described in the section "Optimal ATO Set," for our demonstration using 29 ATOs, the model selected the optimal subset to consist of ATOs 1, 5, 13, 15, 16, 19, 20, 23, 25, 26, and 29. The model output will also include the proper distribution of S&T funds (as shown in Figure 5.4) and remaining lifecycle funds (see Figure 5.7) among 13 FOCs. To develop Figure 5.7, we allocated the funds according to the method described in the section "Remaining S&T Cost Attributable to Each FOC," with RLCC substituted for RSTC.

In sum, if funds are distributed evenly among FOCs, it is likely that some FOC requirements will not be met, while others are unnecessarily over-met. Moreover, one needs guidance from a graph such as Figure 5.7 for an optimal allocation of remaining lifecycle funds among 13 FOCs in order to meet all 13 requirements at the lowest cost.

---

[13] An ATO can also move up in rank by improving its contribution to FOCs. The model can also be used to determine how much more contribution would be needed to move up in rank.

Figure 5.7
Remaining Lifecycle Cost Attributable for Meeting Each of the 13 FOCs

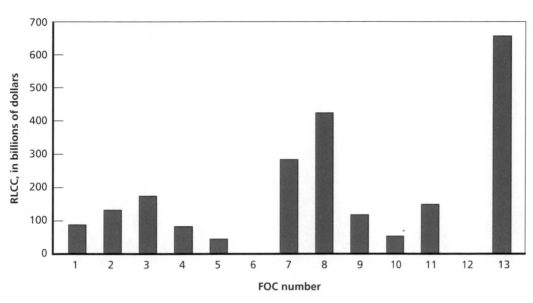

Cost-Effectiveness of Remaining Lifecycle Funds in Developing Individual FOCs

Similarly to Figure 5.5 for the remaining S&T funds, Figure 5.8 shows the remaining lifecycle funds needed to develop 1 percent of capability in each FOC. The cost-effectiveness shown in Figure 5.8 can be used to guide ATO program managers concerning which ATOs to terminate in the event of a cut in the budget for total remaining lifecycle spending. This guide works in the same manner as the one for a cut in budget for total remaining S&T cost that was shown in Figure 5.5. For example, for the same percentage cut in an FOC capability, cutting the FOC 13 capability would yield the largest savings in total remaining lifecycle cost.

Non-Linearity in Total Remaining Lifecycle Cost to Meet FOC Requirements

We want to examine the total remaining lifecycle cost required to meet five levels of possible FOC requirements: 0.8RR, 0.9RR, RR, 1.1RR, and 1.2RR as shown in Figure 5.6. As we did for Figure 5.3, we selected from the pool of 29 ATOs to form an optimal portfolio for each of five FOC requirements and calculated its total remaining lifecycle

Figure 5.8
Some FOCs Need More RLCC Funds to Develop 1 Percent of Their Capabilities

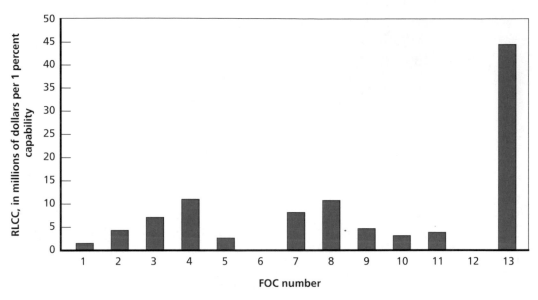

RAND *MG761-5.8*

cost (see Figure 5.9).[14] Figure 5.9 also shows the marginal cost to meet every increment. It takes $2.1 billion to meet the first 80 percent (i.e., 0.8RR). No additional funding is required to meet the next 10-percent increment to reach 0.9RR. The marginal cost to meet the reference requirement is $53 million, followed by $780 million and $830 million to meet the remaining two 10-percent increments up to the 1.2RR level.

The non-linearity is due to the complicated interplay of two factors. First, all else being equal, the model first selects the ATO-derived systems[15] that can meet the requirements most cheaply. Second, because adding an additional ATO can sometimes meet requirements in larger chunks than the 10-percent increment, the cost of meeting the next 10-percent can be lower than that of meeting the current 10-percent increment.[16] The interplay between these two factors results in the fact that the group

---

[14] Meeting 80 percent of the FOC requirements means the model insists that the selected ATOs will meet 80 percent of each of 13 individual FOC requirements. Meeting any other fractions of requirements is defined similarly.

[15] As defined previously, an ATO-derived system is a weapon system or product that is developed under an ATO.

[16] An extreme example can clarify this situation. Suppose there is only one ATO but the same 13 FOC requirements as shown in Figure 5.3 to satisfy. Let us further assume that this ATO can meet 100 percent of every one of 13 FOC requirements at a remaining lifecycle cost of $10 billion. When one asks what the cost is to meet 90 percent of the requirements, the answer is $10 billion. When one asks what additional cost would be required to meet an additional 10 percentage points to reach 100 percent, the answer is zero because meeting the first 90 percent of the requirements has already required the funding of the same ATO and its derived systems. Implicit

Figure 5.9
Remaining Lifecycle Cost Needed to Meet Various Fractions of FOC Requirements

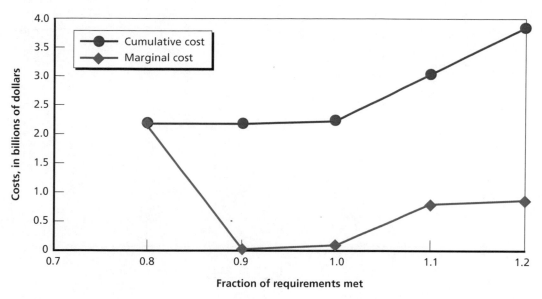

RAND *MG761-5.9*

of ATOs selected by the model to fulfill the 80-percent requirement also fulfills the 90-percent requirement without necessitating any additional funding.

Acquisition planners need to pay attention to this non-linearity of cost when they are asked to estimate the change in remaining lifecycle cost to meet an increase or decrease in FOC requirements. A good decision hinges on knowing not only the change of requirement, but also the change in cost to meet the changed requirements. For example, if the additional lifecycle cost to meet a large increase in FOC requirements is low, the acquisition planners can argue that additional funding is cost-effective. If only a small savings in lifecycle cost would result from a large reduction in FOC requirements, the acquisition planners can likewise make the case that savings should be identified elsewhere, where less loss of future capability will be incurred.

Non-Linearity in Total Remaining S&T Cost to Meet FOC Requirements

The situation with respect to total remaining S&T cost is similar to that described in the previous section for total remaining lifecycle cost. Figure 5.10 shows the non-linearity in total remaining S&T cost to meet various percentage increments of FOC

---

here and in this chapter as a whole, the same number of units of the system derived from this ATO is assumed to be needed to meet the first 90 percent or 100 percent of the requirements.

Figure 5.10
Remaining S&T Cost Needed to Meet Various Fractions of FOC Requirements

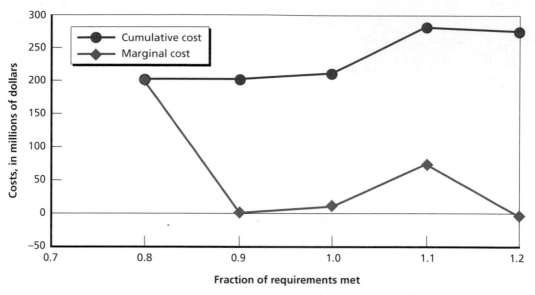

requirements. The cumulative cost is not linear with respect to the percentage increment in requirement, and the marginal cost is not a constant at four increments from 0.8RR to 1.2RR, although they are all 10-percent increments.[17] Again, as with nonlinearity in remaining lifecycle cost, the managers of S&T budgets need to understand the relationship between changes in requirements and changes in S&T resource needs.

Existence of an Optimal Total Remaining S&T Budget

In our formulation, the remaining S&T budget is included as part of the remaining lifecycle cost of the selected ATOs and their derived systems. In this section, we examine the important relationship between the total remaining S&T budget and the total remaining lifecycle cost. Using Figure 5.3 as the starting case, the model selected 11 of the 29 ATOs to yield the lowest total remaining lifecycle cost for meeting all indi-

---

[17] It is possible that the cumulative remaining S&T cost to meet 1.1 times the reference requirement can be higher than that for 1.2 times the reference requirement, because the model minimizes cumulative remaining lifecycle cost, not that of S&T cost. As shown in Figure 5.6, the model picks ATO#4, ATO#27, and others to meet 1.2 × RR, while at 1.1 × RR, the model picks ATO#21 and the same others. ATO#21 has a higher remaining S&T cost than those of ATO#4 and ATO#21 combined but a much lower remaining lifecycle cost.

vidual FOC requirements.[18] The total remaining S&T budget to fund the completion of these 11 selected ATOs is $206 million and the total remaining lifecycle cost is $2.2 billion (see Figure 5.11). Three points are in order. First, for our demonstration case, any amount beyond the $206 million budget would be unnecessary, since there is no need to support additional ATOs whose systems are not needed to meet the requirements.[19] Starting from the optimal funding level (the red point in Figure 5.11), as the total remaining S&T budget increases toward $1 billion along the horizontal axis of Figure 5.11, the MRLCC amount along the vertical axis incorporates this change in total remaining S&T funding. However, under our no-project-failure assumption, these additional S&T funds will be wasted since the 11 selected ATOs already suffice to meet all the FOC requirements. Second, near-term cuts in total remaining S&T funding can lead to drastic increases in lifecycle costs. A reduced total remaining S&T budget is inadequate to support ATOs with high research costs. Thus, the model is forced to select ATOs that need less S&T funds to complete but much higher funding to field their systems. The net result is an increase in total remaining lifecycle cost

Figure 5.11
Existence of an Optimal Remaining S&T Budget

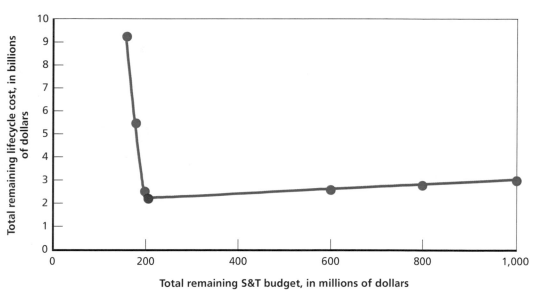

RAND *MG761-5.11*

---

[18] As discussed in the section in which Figure 5.3 appeared, the 11 selected ATOs are 1, 5, 13, 15, 16, 19, 20, 23, 25, 26, and 29.

[19] This statement is based on our assumptions that there is no risk of failing to complete any ATO and fielding its system. As we stated before, this assumption has been relaxed in the ongoing follow-on study. However, an optimal S&T budget still exists even without this assumption.

compared with the optimal solution. In our 29-ATO case, when the total remaining S&T budget is restricted to less than $160 million, the model chooses a portfolio of projects requiring $159 million in total remaining S&T funding. The total remaining lifecycle cost for the projects rises sharply to $9.2 billion from $2.2 billion under the optimal case. In other words, in our demonstration case, saving an estimated $47 million in remaining S&T funds in the near term would cause the Army to pay $7 billion more in lifecycle cost in the long term. This would be a classic example of penny wise and pound foolish. This drastic outcome results from the fact that the total remaining S&T budget is a very small fraction (see Table 4.2) of the total remaining lifecycle cost. Third, there exists an optimal level of total remaining S&T budget, and it is important to make an estimate of it and try hard to avoid cutting the budget below this optimal level. A curve such as that shown in Figure 5.11 would provide the Army with important guidance in the proper allocation of funds between S&T activities and other programs, such as procurement of weapon systems.

Collate Color Text Section

Collate Color Text Section

CHAPTER SIX

Findings and Recommendations

We have developed and demonstrated a method and a linear programming model that account for two important factors in the design and selection of ATOs. These factors are the satisfaction of all individual FOC requirements and the minimization of total remaining lifecycle cost for all of the systems to be deployed. Our goal is to meet all requirements at the lowest total remaining lifecycle cost.

Model Applications and Findings

We demonstrated applications that will ensure that the ATOs selected are the most cost-effective projects to meet the FOC requirements. The applications will also help the Army monitor ATO development to allow early adjustments when the performance and/or lifecycle cost outlook of the project are trending worse than originally anticipated. The following applications of our model should help the Army select and manage ATOs:

- Determination of the extent to which the FOC requirements would be met if all existing ATOs were completed and their systems fielded. This allows the Army to trace the impact of ATOs on FOCs.
- Identification and introduction of new ATOs where existing ATOs leave gaps in meeting FOC requirements.
- Determination of the subset of existing ATOs that can meet all individual FOC requirements at the lowest total remaining lifecycle cost. This determination is of particular interest to the Assistant Secretary of the Army for Acquisition, Logistics, and Technology and the Army Vice Chief of Staff, who co-chair the S&T Advisory Group that provides the final approval of S&T programs.
- Determination of the extent to which each of the individual FOC requirements is exceeded. This provides a safety margin beyond that already embedded in the requirement. The Army, in particular its Training and Doctrine Command, the Assistant Deputy Chief of Staff for Programs (Force Development), and the Army's Research, Development and Engineering Command, needs to know

in which individual FOC requirements these extra safety margins will occur, because these are the requirements for which the Army can have extra capabilities without the need to develop additional costly ATOs and systems.

- Determination of the optimal distribution of funds across FOC requirements. We also found that distributing S&T funds and/or acquisition and operating funds evenly among individual FOC requirements is unlikely to be the least expensive way to meet FOC requirements.

- Determination of which set of ATOs should be ranked high and be protected from budgetary cuts. This set can yield the lowest total remaining lifecycle cost over a range of uncertain future requirements.

- Identification of cost and performance factors that prevent specific ATOs from being ranked higher in importance. Understanding these factors is useful to individual ATO program managers as they develop cost and performance targets. The model determines the specific cost and contribution that the ATO must attain to move to a higher rank.

- Determination of the optimal level of the total remaining S&T budget to meet all of the individual FOC requirements at the lowest total remaining lifecycle cost. The non-linearity in the relationship between the total remaining S&T budget and the total remaining lifecycle cost of materiel systems makes this information critically important as the Army's leadership balances overall Army resource demands. We have found that even small cuts in the total remaining S&T budget can lead to lifecycle costs in later years that are an order of magnitude or two larger than that of the cut. Also, it is possible that the Army can increase the total remaining S&T budget somewhat, which can lead to a much larger savings in lifecycle cost down the road.

The Army and the other services face a persistent challenge of under investment in S&T. For example, in August 2007, John Young, the Pentagon's DDR&E, argued that the S&T budget should be 3 percent, instead of the current 2.2 percent, of the DoD total budget. This 3-percent guideline came from a 1998 Defense Science Board study that found that companies such as pharmaceutical giants regularly invest this percentage of their budgets in scientific research ("Young: Pentagon Should Invest Billions More in Science, Technology," 2007). Skeptics could argue that the military is different from commercial companies and the 3-percent rule is not applicable.

Our study offers a direct way to quantify the optimal total remaining S&T budget as a percentage of the total remaining lifecycle cost. The same method can be used to find the optimal (total) S&T budget as a percentage of the (total) lifecycle cost by simply replacing the total remaining S&T and lifecycle budgets with the total S&T

and lifecycle budget.[1] While our model is currently in the demonstration stage for the Army, it can be generalized to include all of the defense S&T programs. This generalization would follow the same approach of asking what the optimal DoD S&T budget should be in order to meet all of the Pentagon's individual FOC and other requirements being addressed by S&T projects at the lowest total cost. This would address the question of whether an S&T budget as 3 percent of the total budget is optimal.

Iterative Improvements on Value and Cost Estimates

Meeting the FOC requirements on an individual basis, incorporating the S&T budget constraint, and using lowest remaining lifecycle cost as the objective function are three major aspects of our method and model that are relatively non-controversial, as is our use of a linear programming model. However, some might argue that the components of remaining lifecycle cost, especially the unit purchase cost of the system and the number of units to be fielded, are unknown at the S&T stage.

We offer two points to counter such an argument. First, when an Army ATO reaches the 6.3 program stage, this stage is equivalent to the CR and the TD stage in the DoD's Defense Acquisition Management Framework under its new acquisition policy. If the CR/TD is for an ACAT I or IA program, DoD requires an AoA, which includes an estimate of all of the cost components discussed in this monograph. We argue that if one can determine the cost of a major system at its CR/TD stage, one can similarly determine that of a system derivable from an Army ATO, because both developments are at the same stage. Besides, we have estimated that 11 of the 29 ATOs can lead to acquisition programs that require an AoA. For these systems, DoD expects that their lifecycle costs can be estimated at the CR/TD stage, which we argue corresponds to 6.3 in the S&T phase. Similarly, if one can estimate the lifecycle cost for the major systems at their 6.3 stage, one should be able to do the same for the non-major systems. Second, the fact that the estimates are subject to large uncertainties is no reason for one to avoid incorporating lifecycle cost into the consideration of S&T project selection and portfolio management, since keeping the lifecycle cost affordable is a key objective

---

[1]   Since we are focusing on ATOs, which are established during 6.2 and 6.3 programs, our total remaining S&T budget does not include funds for 6.1 programs. However, one can total the S&T budgets for 6.1, 6.2, and 6.3 and use our method and model to determine the optimal S&T budget as a percentage of the total lifecycle cost. This would address whether 3 percent is optimal.

of many S&T projects.[2] Instead, one should incorporate both the estimates and their uncertainties into the analysis.[3]

We propose an iterative procedure to improve the estimates of value (contribution to FOC requirements) and lifecycle cost. The Army S&T community has an extensive technology planning, review, and oversight process. At the beginning of each fiscal year, the Deputy Assistant Secretary for Research and Technology and the Director of the Force Development Office of the Deputy Chief of Staff for Programs issue guidance on new technology proposals and existing technology reviews (U.S. Department of Defense, 2002, pp. 4–5). New technology proposals are reviewed and approved during the Army's annual S&T oversight process. This process also reviews ongoing technology projects—including ATOs, ATDs, and ACTDs (which have been in development for three years)—that have been revised in terms of cost, schedule, or scope, and that have been completed. The Army S&T Programs receive management direction and approval from three executive-level groups. The first level of review is the Warfighting Technical Council (WTC), a one-star-level group that performs detailed reviews of all proposed and ongoing S&T programs, including ATOs. The second level of review is the Army S&T Working Group, a two-star-level resolution of issues that reviews and approves S&T efforts. The third level, providing the final approval, is the S&T Advisory Group, a four-star-level group.

We envision the following procedure for the Army S&T community to develop, incorporate, and update lifecycle cost into its review and approval process of ATOs.[4] During the first four months of a fiscal year, the WTC estimates the EVs of the ATO's contributions to FOC requirements and the components of the lifecycle cost according to the estimation procedure proposed in this study. This estimation will also provide target values and costs for an ATO to meet. These data would be delivered to the relevant ATO program managers by January 31.

During the next four months, the management and the engineers of the pertinent ATO would be required to validate the data from the WTC against their own engineering estimates. For each estimate, they would have to state agreement or disagreement. In the latter case, they would be required to provide their own estimate and the

---

[2]   We recognize that a small number of S&T projects even at the 6.3 program stage could continue to be broadly focused on the advancement of technologies in general and with no application in mind to particular systems. However, the capability contribution and the lifecycle cost of many of the even broadly focused ATOs can be measured and included in our model and analysis. For those that cannot be measured, a separate S&T budget can be set aside for them, and they can be excluded from the general analysis proposed in this study. In this study, we excluded one such ATO, as discussed in the section "ATOs Selected for Our Analysis" in Chapter Four. It could be argued that our general analysis should be carried out for the majority of the S&T projects at the 6.3 stage.

[3]   Incorporation of the uncertainties is being pursued in our ongoing follow-on study.

[4]   We suggest an initial focus on ATOs under 6.3 programs, since they are more mature and their capability and lifecycle cost are better known. Then, one should attempt to include ATOs under 6.2 programs, and finally 6.2 and 6.3 programs not selected for ATOs.

reasons why their own is more justified and why the target values and costs cannot be met. Their report would be sent to the WTC by May 31.

During the last four months of the fiscal year, the WTC would consider the ATO report, but arrive at its own assessment about the merits and potential problems of the ATO. It would then suggest a course of action aimed at resolving these problems. On the ATO estimates with which it does not agree, it would explain the reason for non-acceptance. This process would be repeated annually for ongoing ATOs in order to develop and incorporate changes in FOC requirements and updates in ATO program estimates regarding their contributions to FOCs and their remaining lifecycle costs.

We anticipate that the exchanges between the WTC and the ATO personnel will improve the capability and lifecycle cost estimates, allowing the goal of fielding the most cost-effective systems to be factored into the design and development of the S&T programs. These exchanges will alert the WTC and ATO managers of potential problems and allow them to take corrective actions in a timely manner.

Recommendations

We recommend three actions for designing, funding, and managing the Army's S&T portfolio so as to meet all of the Army's FOC requirements at the lowest total remaining lifecycle cost:

1.   Establish a pilot program to test the practicality and usefulness of the iterative procedure described above for better estimating an ATO's contributions to individual FOC requirements and its derived system's lifecycle cost. This pilot program, if successful, will be an important step toward the goal of meeting future capabilities at the lowest cost.
2.   Set up a pilot program to carry out the method, model, and estimation procedures proposed in this monograph. The pilot program can be based on existing ATOs and should include all the applications suggested above in this chapter, which are elaborated in Chapter Five, plus the use of the much simpler Delphi method for estimation of rough cost levels as discussed in Appendix F.
3.   Since the method and model developed here are applicable to the portfolio management of Army programs at other than the S&T stage, apply the approach here to the selection and management of programs in the SDD stage. Other services and DoD may wish to try this approach as well.

In an ongoing follow-on study, we are expanding our method to incorporate uncertainties in S&T project successes, with the objective of making the model a better reflection of the uncertain future. The ongoing study examines the impacts of uncertainties on issues such as the selection of ATOs that can meet FOC requirements over a wide range of uncertain futures. In other words, we want to describe how to select a portfolio of ATOs that is robust against uncertainties.

We believe that not only the Army, but also the other services and DoD as a whole need to consider lifecycle cost at an early stage in the development of weapon and other systems in order to allow adjustments, where necessary, to achieve affordable systems that meet all individual capability requirements.

Additional Information on Army Acquisition Categories and Combat Developer

This appendix describes the primary criteria for classifying an Army ACAT. It also provides additional information about the role of a combat developer.

Classification Criteria for ACATs

Defense acquisition programs are classified into distinct ACATs that determine the level of the MDA (U.S. Army, 2003). ACAT I programs are MDAPs and have two subcategories: ACAT ID and ACAT IC. ACAT ID programs are MDAPs whose MDA is the Under Secretary of Defense for Acquisition, Technology and Logistics. The primary criteria for classifying a program as an ACAT ID are RDT&E costs of more than $365 million in FY 2000 constant dollars or procurement costs of more than $2.19 billion.[1] ACAT IC programs are MDAPs whose MDA is the service acquisition executive. In this Army study, the service acquisition executive is the AAE. Its primary criteria are the same as those for ACAT ID.

ACAT IA programs are MAISs or programs. ACAT IA has two subcategories: ACAT IAM and ACAT IAC. The former programs are MAISs whose MDA is the DoD Chief Information Officer, while the latter is the AAE. ACAT IA programs are those with cost exceeding $32 million in a single year or $126 million in total program cost or $378 million in total lifecycle cost.

ACAT II programs are those programs that do not meet the criteria for an ACAT I program (i.e., ID and IC), but are major systems or are designated as ACAT II by the AAE. An ACAT II program is one that has more than $140 million in RDT&E cost or more than $660 million in procurement cost.

ACAT III programs are those that do not meet the criteria for ACAT I (i.e., ID and IC), ACAT IA, or ACAT II. The MDA is designated by the AAE. They are non-

---

[1] All dollar figures in this section are FY 2000 constant dollars.

major systems, including command, control, communications, and computers (C4) and information technology. There are no fiscal criteria for this category.

The Army eliminated the designation of ACAT IV programs in an Army acquisition policy memorandum issued on December 31, 2003.

Role of a Combat Developer

While TRADOC usually tasks TRAC to conduct AoAs for ACAT I, ACAT IA and ACAT II programs, CBTDEV is responsible for conducting the remaining ACAT II and III program AoAs if required by the MDA. To describe the combat developer, we simply quote a passage from the Army Acquisition Procedure:

> The combat developer is that command, organizational element (including base operations and HQDA [Headquarters, Department of the Army]), and individual responsible for preparing and processing the materiel requirement document (MRD) and representing the user (organization and individual) of the new or modified system throughout the acquisition process. Combat developers apply to both materiel systems and information technology systems. Assignments of branch or specified proponent under AR 5-22 (as is the case for commands such as TRADOC, Medical Command (MEDCOM), Space and Missile Defense Command (SMDC), and Intelligence and Security Command (INSCOM)) bring with it combat development responsibilities for deployable and non-deployable materiel and information warfighting systems. The Materiel Developer (MATDEV), in coordination with the Combat Developer (CBTDEV) and Training Developer (TNGDEV), performs concept studies on the best technological candidates identified by the technology trade-offs conducted during the Determination of Mission Need phase. These studies develop rough performance estimates and research, development, and acquisition (RDA) cost estimates with sufficient resolution to permit trade-offs among system performance, operational capability, requirements, and costs. Concept studies identify system concept alternatives for the AoA, provide input for development of the program baseline, and influence the ORD [operational requirement document] through interaction with the CBTDEV requirements analyses (U.S. Army, 1999, p. 16).

Whenever an AoA is not required for an ACAT III program, the CBTDEV keeps records of the materiel need determination process, the requirements analysis, and the operational analysis to provide the analytic basis for the Capability Development

Document (CDD)[2] that is approved at Milestone B and for the Capability Production Document (CPD)[3] approved at Milestone C.

---

[2] The CDD describes what the SDD effort will be and provides the key performance parameters for the increment during the SDD phase. It also describes the program needed to attain a complete solution.

[3] The CPD describes the PD effort to produce the materiel solution and provides key performance parameters for the production increment (Center for Program Management, 2003, slide 26).

APPENDIX B
Estimation of Expected Values[1]

This appendix provides information on how we estimated the contributions of ATOs to FOCs listed in Table 4.1. As in the example of Overwatch (ATO#3) discussed in Chapter Four, we use the following data sources:

- The July 2005 Army S&T Master Plan, which was the latest version available during the time of our study.
- 2006 DTOs, which were the latest available during the time of our study. DTOs are updated regularly on the Web by DDR&E. We selected ATOs that are also DTOs for this study, because more data are available on DTOs.

As described in Chapter Three, we use a slightly modified version of the FOCs described in TRADOC pamphlet 525-66 (U.S. Army, 2005b). Based on the data on objectives, metrics, payoffs, challenges, and milestones in the data sources listed above, and the definitions of the FOCs, we list below our conclusions as to the FOC capability requirements which each ATO contributes to meeting, and then present our estimates of CV for each ATO. The CV estimation method follows that illustrated in Chapter Four for the Overwatch ACTD (ATO#3), using the FOC requirement gap space coverage estimation approach described in Chapter Three, the situations of Figure 3.2, and the categories of Table 3.1. As noted above, CV is presented, rather than a precise EV percentage, for each ATO. The ATO was apportioned a value based on the number of gaps it helps fill times the percentage of particular FOC sub-requirements it seeks to address. To correctly convert CV to EV, we would have also had to determine, in percentage terms, how well the ATO met each gap and FOC sub-requirement. Rather than performing this task, we simply reduced the CV percentage calculated by 50 per-

---

[1] It should be emphasized that this study focused on methodology development. To demonstrate the methodology, we took a retrospective look at how well a small subset of the 2005 ATO projects could meet a hypothetical set of gaps, which we assumed occurred in every FOC requirement and sub-requirement. In other words, the study did not use data on real capability gaps and provided no information on how well the Army S&T portfolio meets the actual gaps. Moreover, since the ATOs were evaluated against hypothetical gaps, one should not draw any conclusions about the merits or drawbacks of any specific S&T project discussed in this study in meeting real Army capability gaps.

cent. This is a rough approximation that does not interfere with the demonstration of the model. Further refinements can be made to the methodology in a full-scale pilot program.

ATO#1: Small Unit Operations

The objective of this ATO is to demonstrate a Soldier Radio Waveform network communications capability in restrictive environments. It contributes to meeting capability requirements in the following FOCs: FOC 1, battle command; FOC 2, battlespace awareness; FOC 3, mounted/dismounted maneuver; FOC 6, maneuver support; and FOC 13, human engineering.

Since the Soldier Radio Waveform network communications capability supports the foot soldier, according to Figure 3.2 it applies to situations (2) and (3), on the way to the battlefield and on the battlefield. Therefore, we multiply its gap coverage estimates by two-thirds. The following describes our estimates of its CV contributions to each of the FOCs listed above.

FOC 1: Battle Command

Of the four categories in Table 3.1, ATO#1 applies only to "communications," so our gap coverage estimate is one-fourth. According to TRADOC pamphlet 525-66 (U.S. Army, 2005b), FOC 1 has six sub-requirements, and ATO#1 contributes to two of these, "layered, integrated C2 for joint, multinational, interagency operations on the move" and "networked force optimized for mobile operations." Therefore, we multiply our gap coverage estimate by one-third. Thus, our estimate of the CV contribution of ATO#1 to FOC 1 ($CV_{1,1}$) is

$$CV_{1,1} = 2/3 \times 1/4 \times 1/3 = 0.0556.$$

FOC 2: Battlespace Awareness

Because ATO#1 is a communications device, its contribution to battlespace awareness is exclusively in the sub-requirement "C2 of battlespace awareness assets." The appropriate categories in Table 3.1 are thus the categories listed under battle command, for which ATO#1 provides only one of four (communications). Therefore our gap coverage estimate is one-fourth. According to TRADOC pamphlet 525-66 (U.S. Army, 2005b), FOC 2 has six sub-requirements, and ATO#1 contributes to only one of these. Therefore, we multiply our gap coverage estimate by one-sixth. Thus, our estimate of the CV contribution of ATO#1 to FOC 2 ($CV_{1,2}$) is

$$CV_{1,2} = 2/3 \times 1/4 \times 1/6 = 0.0278.$$

FOC 3: Mounted-Dismounted Maneuver

Of the four categories in Table 3.1, ATO#1 applies only to "equipment and supplies," so our gap coverage estimate is one-fourth. According to TRADOC pamphlet 525-66 (U.S. Army, 2005b), FOC 3 has ten sub-requirements, and ATO#1 contributes to two of these, "sustainment with minimal load and logistics footprint" and "operations in urban and complex terrain." Therefore, we multiply our gap coverage estimate by one-fifth. Thus, our estimate of the CV contribution of ATO#1 to FOC 3 ($CV_{1,3}$) is

$$CV_{1,3} = 2/3 \times 1/4 \times 1/5 = 0.0333.$$

FOC 6: Maneuver Support

Of the four categories in Table 3.1, ATO#1 applies only to "equipment and supplies," so our gap coverage estimate is one-fourth. According to TRADOC pamphlet 525-66 (U.S. Army, 2005b), FOC 6 has seven sub-requirements, and ATO#1 contributes to only one of these, "understand battlespace environment." Therefore, we multiply our gap coverage estimate by one-seventh. Thus, our estimate of the CV contribution of ATO#1 to FOC 6 ($CV_{1,6}$) is

$$CV_{1,6} = 2/3 \times 1/4 \times 1/7 = 0.0238.$$

FOC 13: Human Engineering

Of the three categories in Table 3.1, ATO#1 applies only to engineering of "systems," so our gap coverage estimate is one-third. According to TRADOC pamphlet 525-66 (U.S. Army, 2005b), FOC 13 has four sub-requirements and ATO#1 contributes only to "reduce soldier dismounted movement approach load." However, ATO#1 only contributes to one of the three categories of soldier load defined in the notes of Table 3.1. Therefore, we multiply our gap coverage estimate by one-fourth and also by one-third. Thus, our estimate of the CV contribution of ATO#1 to FOC 13 ($CV_{1,13}$) is

$$CV_{1,13} = 2/3 \times 1/3 \times 1/4 \times 1/3 = 0.0185.$$

ATO#2: Vaccines for Prevention of Malaria

The objective of this ATO is to develop vaccines to reduce the incidence of malaria in deployed forces by 80 percent. It contributes to meeting capability requirements in FOC 7, personnel protection.

Since malaria can be contracted anywhere, ATO#2 applies to all three of the situations of Figure 3.2, so there is no multiplier. Of the two categories in Table 3.1,

ATO#2 applies only to "disease," so our gap coverage estimate is one-half. Moreover, malaria is only one of the five categories for disease defined in the notes of Table 3.1, so we multiply by an additional factor of one-fifth. There are no sub-requirements for this FOC, but we introduce an additional factor of one-half to provide redundancy for protection of the warfighter. Thus, our estimate of the CV contribution of ATO#2 to FOC 7 ($CV_{2,7}$) is

$$CV_{2,7} = 1/2 \times 1/5 \times 1/2 = 0.0500.$$

ATO#3: Overwatch ACTD

This ATO is the example discussed in Chapter Four.

ATO#4: Unmanned Ground Mobility

The objective of this ATO is to demonstrate unmanned ground mobility technologies that support a range of ground and amphibious missions. It contributes to meeting capability requirements in the following FOCs: FOC 6, maneuver support; FOC 7, personnel protection; FOC 8, asset protection; and FOC 11, maneuver sustainment.

Since ATO#4 supports ground and amphibious missions, according to Figure 3.2 it applies to situations (2) and (3), on the way to the battlefield and on the battlefield. Therefore, we multiply its gap coverage estimates by two-thirds. The following describes our estimates of the CV contributions of ATO#4 to each of the FOCs listed above.

FOC 6: Maneuver Support

Of the four categories in Table 3.1, ATO#4 applies to "forces," "mobility," and "weapons," so our gap coverage estimate is three-fourths. According to TRADOC pamphlet 525-66 (U.S. Army, 2005b), FOC 6 has seven sub-requirements, and ATO#4 contributes to only one of these, "assure mobility." Accordingly, our estimate of $CV_{4,6}$ is

$$CV_{4,6} = 2/3 \times 3/4 \times 1/7 = 0.0714.$$

FOC 7: Personnel Protection

Of the two categories in Table 3.1, ATO#4 applies only to "injury," so our gap coverage estimate is one-half. Moreover, of the four categories for injury defined in the notes of Table 3.1, ATO#4 applies to "LOS fire" and "hazards," so we multiply by an additional factor of one-half. There are no sub-requirements for this FOC, but we introduce an additional factor of one-half to provide redundancy for protection of the warfighter. Accordingly, our estimate of $CV_{4,7}$ is

$$CV_{4,7} = 2/3 \times 1/2 \times 1/2 \times 1/2 = 0.0833.$$

FOC 8: Asset Protection

Of the four categories in Table 3.1, ATO#4 applies to "LOS fire" and "hazards," so our gap coverage estimate is one-half. There are no sub-requirements for this FOC, but we introduce an additional factor of one-half to provide redundancy for protection of warfighting assets. Accordingly, our estimate of $CV_{4,8}$ is

$$CV_{4,8} = 2/3 \times 1/2 \times 1/2 = 0.167.$$

FOC 11: Maneuver Sustainment

Of the four categories in Table 3.1, ATO#4 applies to "forces," "mobility," and "weapons," so our gap coverage estimate is three-fourths. According to TRADOC pamphlet 525-66 (U.S. Army, 2005b), FOC 11 has nine sub-requirements, and ATO#4 contributes to two of these, "improved sustainability" and "enhancements in readiness, reliability, maintainability, and commonality for sustained operational tempo." Accordingly, our estimate of $CV_{4,11}$ is

$$CV_{4,11} = 2/3 \times 3/4 \times 2/9 = 0.111.$$

ATO#5: Warfighter-System Interaction

The objective of this ATO is to provide effective interfaces for both the mounted and dismounted warfighters interacting with their platforms as well as the unmanned ground and air assets they control. It contributes to meeting capability requirements in the following FOCs: FOC 1, battle command; FOC 2, battlespace awareness; FOC 3, mounted/dismounted maneuver; FOC 7, personnel protection; FOC 8, asset protection; FOC 11, maneuver sustainment; and FOC 13, human engineering.

Since ATO#5 supports the warfighter, according to Figure 3.2 it applies to situations (2) and (3), on the way to the battlefield and on the battlefield. Therefore, we multiply its gap coverage estimates by two-thirds. The following describes our estimates of its CV contributions to each of the FOCs listed above.

FOC 1: Battle Command

Of the four categories in Table 3.1, ATO#5 applies to all, so there is no multiplier. According to TRADOC pamphlet 525-66 (U.S. Army, 2005b), FOC 1 has six sub-requirements, and ATO#1 contributes to only one of these, "layered, integrated C2 for joint, multinational, interagency operations on the move." Therefore, we multiply our gap coverage estimate by one-sixth. Thus, our estimate of the CV contribution of ATO#5 to FOC 1 ($CV_{5,1}$) is

$$CV_{5,1} = 2/3 \times 1/6 = 0.111.$$

FOC 2: Battlespace Awareness

Of the four categories in Table 3.1, ATO#5 applies to "LOS fire," "enemy force location," and "hazards." Therefore, our gap coverage estimate is three-fourths. According to TRADOC pamphlet 525-66 (U.S. Army, 2005b), FOC 2 has six sub-requirements, and ATO#1 contributes to only one of these, "C2 of battlespace awareness assets." Therefore, we multiply our gap coverage estimate by one-sixth. Thus, our estimate of the CV contribution of ATO#5 to FOC 2 ($CV_{5,2}$) is

$$CV_{5,2} = 2/3 \times 3/4 \times 1/6 = 0.0833.$$

FOC 3: Mounted-Dismounted Maneuver

Of the four categories in Table 3.1, ATO#5 applies to "forces," "mobility," and "weapons," so our gap coverage estimate is three-fourths. According to TRADOC pamphlet 525-66 (U.S. Army, 2005b), FOC 3 has ten sub-requirements, and ATO#5 contributes to two of these, "effective battle command" and "human engineering for improved soldier-system interface." Therefore, we multiply our gap coverage estimate by one-fifth. Thus, our estimate of the CV contribution of ATO#5 to FOC 3 ($CV_{5,3}$) is

$$CV_{5,3} = 2/3 \times 3/4 \times 1/5 = 0.100.$$

FOC 7: Personnel Protection

Of the two categories in Table 3.1, ATO#5 applies only to "injury," so our gap coverage estimate is one-half. There are no sub-requirements for this FOC, but we introduce an additional factor of one-half to provide redundancy for protection of the warfighter. Accordingly, our estimate of $CV_{5,7}$ is

$$CV_{5,7} = 2/3 \times 1/2 \times 1/2 = 0.167.$$

FOC 8: Asset Protection

Of the four categories in Table 3.1, ATO#5 applies to all, so there is no multiplier. There are no sub-requirements for this FOC, but we introduce an additional factor of one-half to provide redundancy for protection of warfighting assets. Accordingly, our estimate of $CV_{5,8}$ is

$$CV_{5,8} = 2/3 \times 1/2 = 0.333.$$

FOC 11: Maneuver Sustainment

Of the four categories in Table 3.1, ATO#5 applies to "forces," "mobility," and "weapons," so our gap coverage estimate is three-fourths. According to TRADOC pamphlet 525-66 (U.S. Army, 2005b), FOC 11 has nine sub-requirements, and ATO#5 contributes to only one of these, "improved soldier support." Accordingly, our estimate of $CV_{5,11}$ is

$$CV_{5,11} = 2/3 \times 3/4 \times 1/9 = 0.0556.$$

FOC 13: Human Engineering

Of the three categories in Table 3.1, ATO#5 applies to engineering of the "people or tasks" and "human-system interfaces," so our gap coverage estimate is two-thirds. According to TRADOC pamphlet 525-66 (U.S. Army, 2005b), FOC 13 has four sub-requirements and ATO#5 contributes to two of these, "decrease task complexity" and "exploit unmanned technology in manned systems." Therefore, we multiply our gap coverage estimate by one-half. Thus, our estimate of the CV contribution of ATO#5 to FOC 13 ($CV_{5,13}$) is

$$CV_{5,13} = 2/3 \times 2/3 \times 1/2 = 0.222.$$

ATO#6: Advanced Rotorcraft Technologies

The objective of this ATO is to explore promising new classes of rotorcraft suitable for both manned and unmanned warfare. It contributes to meeting capability requirements in the following FOCs: FOC 3, mounted/dismounted maneuver; and FOC 11, maneuver sustainment.

Since ATO#6 principally provides "focused logistics," according to Figure 3.2 it applies to situation (2), on the way to the battlefield. Therefore, we multiply its gap coverage estimates by one-third. The following describes our estimates of its CV contributions to each of the FOCs listed above.

FOC 3: Mounted-Dismounted Maneuver

Of the four categories in Table 3.1, ATO#6 can provide all, so there is no multiplier. According to TRADOC pamphlet 525-66 (U.S. Army, 2005b), FOC 3 has ten sub-requirements, and ATO#5 contributes to two of these, "ability to deploy rapidly" and "sustainment with minimal logistics footprint." Therefore, we multiply our gap coverage estimate by one-fifth. Thus, our estimate of the CV contribution of ATO#6 to FOC 3 ($CV_{6,3}$) is

$$CV_{6,3} = 1/3 \times 1/5 = 0.0667.$$

FOC 11: Maneuver Sustainment

Of the four categories in Table 3.1, ATO#6 can provide all, so there is no multiplier. According to TRADOC pamphlet 525-66 (U.S. Army, 2005b), FOC 11 has nine sub-requirements, and ATO#6 contributes to two of these, "improved sustainability" and "enhancements in readiness, reliability, maintainability, and commonality for sustained operational tempo." Accordingly, our estimate of $CV_{6,11}$ is

$$CV_{6,11} = 1/3 \times 2/9 = 0.0741.$$

ATO#7: Theater Effects-Based Operations ACTD

The objective of this ATO is to operationalize U.S. Joint Forces Command's effects-based operations (EBO) concepts within United States Forces, Korea (USFK) through technology and developing EBO concepts unique to the USFK operational theater. It contributes to meeting capability requirements in the following FOCs: FOC 1, battle command; FOC 2, battlespace awareness; FOC 3, mounted/dismounted maneuver; and FOC 13, human engineering.

Since ATO#7 involves "C2 of net-centric operations," according to Figure 3.2 it applies to situation (1), off the battlefield. Therefore, we multiply its gap coverage estimates by one-third. The following describes our estimates of its CV contributions to each of the FOCs listed above.

FOC 1: Battle Command

Of the four categories in Table 3.1, ATO#7 applies to all, so there is no multiplier. According to TRADOC pamphlet 525-66 (U.S. Army, 2005b), FOC 1 has six sub-requirements, and ATO#7 contributes to three of these, "layered, integrated C2 for joint, multinational, interagency operations on the move," "decision planning and support capabilities," and "information operations integrated with information management and ISR." Therefore, we multiply our gap coverage estimate by one-half. Thus, our estimate of the CV contribution of ATO#7 to FOC 1 ($CV_{7,1}$) is

$$CV_{7,1} = 1/3 \times 1/2 = 0.167.$$

FOC 2: Battlespace Awareness

Of the four categories in Table 3.1, ATO#7 applies to all, so there is no multiplier. According to TRADOC pamphlet 525-66 (U.S. Army, 2005b), FOC 2 has six sub-requirements, and ATO#7 contributes to two of these, "ability to manage knowledge" and "fusion of information." Therefore, we multiply our gap coverage estimate by one-third. Thus, our estimate of the CV contribution of ATO#7 to FOC 2 ($CV_{7,2}$) is

$$CV_{7,2} = 1/3 \times 1/3 = 0.111.$$

FOC 3: Mounted-Dismounted Maneuver

Of the four categories in Table 3.1, ATO#7 applies to all, so there is no multiplier. According to TRADOC pamphlet 525-66 (U.S. Army, 2005b), FOC 3 has ten sub-requirements, and ATO#7 contributes to only one of these, "effective battle command." Therefore, we multiply our gap coverage estimate by one-tenth. Thus, our estimate of the CV contribution of ATO#7 to FOC 3 ($CV_{7,3}$) is

$$CV_{7,3} = 1/3 \times 1/10 = 0.0333.$$

FOC 13: Human Engineering

Of the three categories in Table 3.1, ATO#5 applies to "engineering of the systems," so our gap coverage estimate is one-third. According to TRADOC pamphlet 525-66 (U.S. Army, 2005b), FOC 13 has four sub-requirements and ATO#7 contributes to only one of these, "decrease task complexity." Therefore, we multiply our gap coverage estimate by one-fourth. Thus, our estimate of the CV contribution of ATO#7 to FOC 13 ($CV_{7,13}$) is

$$CV_{7,13} = 1/3 \times 1/3 \times 1/4 = 0.0278.$$

ATO#8: Power and Energy

The objective of this ATO is to develop and demonstrate the necessary enabling technologies for an advanced hybrid-electric FCS ground vehicle accommodating pulse power for lethality and survivability systems. It contributes to meeting capability requirements in FOC 3, mounted-dismounted maneuver; and FOC 11, maneuver sustainment.

Since ATO#8 provides power and energy for combat systems, according to Figure 3.2 it applies to situations (2) and (3), on the way to the battlefield and on the battlefield. Therefore, we multiply its gap coverage estimates by two-thirds. The following describes our estimates of its CV contributions to each of the FOCs listed above.

FOC 3: Mounted-Dismounted Maneuver

Of the four categories in Table 3.1, ATO#8 provides power and energy for all, so there is no multiplier. According to TRADOC pamphlet 525-66 (U.S. Army, 2005b), FOC 3 has ten sub-requirements, and ATO#8 contributes to only one of these, "Sustainment with minimal load and logistics footprint." Therefore, we multiply our gap coverage estimate by one-tenth. Thus, our estimate of the CV contribution of ATO#8 to FOC 3 ($CV_{8,3}$) is

$$CV_{8,3} = 2/3 \times 1/10 = 0.0667.$$

FOC 11: Maneuver Sustainment

Of the four categories in Table 3.1, ATO#8 provides power and energy for all, so there is no multiplier. According to TRADOC pamphlet 525-66 (U.S. Army, 2005b), FOC 11 has nine sub-requirements, and ATO#8 contributes to two of these, "improved sustainability" and "power and energy." Accordingly, our estimate of $CV_{8,11}$ is

$$CV_{8,11} = 2/3 \times 2/9 = 0.148.$$

ATO#9: Future Tactical Truck System ACTD

The objective of this ATO is to assess key technologies and emerging Future Army Sustainment concepts in developing the requirements of an optimized distribution platform as well as the command and control platform for the unit of action. It contributes to meeting capability requirements in FOC 6, maneuver support; and FOC 11, maneuver sustainment.

Since ATO#9 provides sustainment for combat systems, according to Figure 3.2 it applies to situations (2) and (3), on the way to the battlefield and on the battlefield. Therefore, we multiply its gap coverage estimates by two-thirds. The following describes our estimates of its CV contributions to each of the FOCs listed above.

FOC 6: Maneuver Support

Of the four categories in Table 3.1, ATO#9 applies to "mobility," "weapons," and "equipment and supplies," so our gap coverage estimate is three-fourths. According to TRADOC pamphlet 525-66 (U.S. Army, 2005b), FOC 6 has seven sub-requirements, and ATO#9 contributes to only one of these, "assure mobility." Accordingly, our estimate of $CV_{9,6}$ is

$$CV_{9,6} = 2/3 \times 3/4 \times 1/7 = 0.0714.$$

FOC 11: Maneuver Sustainment

Of the four categories in Table 3.1, ATO#9 applies to "mobility," "weapons," and "equipment and supplies," so our gap coverage estimate is three-fourths. According to TRADOC pamphlet 525-66 (U.S. Army, 2005b), FOC 11 has nine sub-requirements, and ATO#9 contributes to three of these, "improved sustainability," "power and energy," and "enhancements in readiness, reliability, maintainability, and commonality for sustained operational tempo." Accordingly, our estimate of $CV_{9,11}$ is

$$CV_{9,11} = 2/3 \times 3/4 \times 1/3 = 0.167.$$

ATO#10: Joint Precision Airdrop System ACTD

The objective of this ATO is to provide a fast, flexible, direct projection-based distribution system to facilitate rapid strategic and tactical deployment of the Future Force and just-in-time resupply to most locations throughout the world. It contributes to meeting capability requirements in FOC 10, strategic responsiveness and deployability; and FOC 11, maneuver sustainment.

Since ATO#10 provides deployment, according to Figure 3.2 it applies to situations (2) and (3), on the way to the battlefield and on the battlefield. Therefore, we multiply its gap coverage estimates by two-thirds. The following describes our estimates of its CV contributions to each of the FOCs listed above.

FOC 10: Strategic Responsiveness and Deployability

Of the three categories in Table 3.1, ATO#10 provides "transport" and "delivery," so our gap coverage estimate is two-thirds. According to TRADOC pamphlet 525-66 (U.S. Army, 2005b), FOC 10 has two sub-requirements, and ATO#10 applies to both, so there is no multiplier. Accordingly, our estimate of $CV_{10,10}$ is

$$CV_{10,10} = 2/3 \times 2/3 = 0.444.$$

FOC 11: Maneuver Sustainment

Of the four categories in Table 3.1, ATO#10 applies to all, so there is no multiplier. According to TRADOC pamphlet 525-66 (U.S. Army, 2005b), FOC 11 has nine sub-requirements, and ATO#10 contributes to two of these, "improved sustainability" and "global precision delivery enhancements." Accordingly, our estimate of $CV_{10,11}$ is

$$CV_{10,11} = 2/3 \times 2/9 = 0.148.$$

ATO#11: Warfighter Physiological Status Monitoring

The objective of this ATO is to develop an adaptable, modular suite of physiological sensors with supporting algorithms for multiple service use/application. It contributes to meeting capability requirements in FOC 11, maneuver sustainment.

Since its sensors have multiple uses/applications, according to Figure 3.2 ATO#11 applies to all three of the situations, off the battlefield, on the way to the battlefield, and on the battlefield. Therefore, there is no multiplier. The following describes our estimates of its CV contributions to each of the FOCs listed above.

FOC 11: Maneuver Sustainment

Of the four categories in Table 3.1, ATO#11 applies only to "forces," so our gap coverage estimate is one-fourth. According to TRADOC pamphlet 525-66 (U.S. Army, 2005b), FOC 11 has nine sub-requirements, and ATO#11 contributes to four of these, "improved sustainability," "enhancements in readiness, reliability, maintainability, and commonality for sustained operational tempo," "global force health and fitness," and "global casualty care management." Accordingly, our estimate of $CV_{11,11}$ is

$$CV_{11,11} = 1/4 \times 4/9 = 0.111.$$

ATO#12: Battlefield Treatment of Fractures and Soft Tissue Trauma Care

The objective of this ATO is to develop or improve cost-effective methods and devices to reduce the long-term adverse effects of injuries and to reduce mission and medical impact associated with potentially extended evacuation times. It contributes to meeting capability requirements in FOC 11, maneuver sustainment.

Since it is battlefield treatment, according to Figure 3.2 ATO#12 applies only to situation (3), on the battlefield. Therefore, we multiply its gap coverage estimates by one-third. The following describes our estimates of its CV contributions to each of the FOCs listed above.

FOC 11: Maneuver Sustainment

Of the four categories in Table 3.1, ATO#12 applies only to "forces," so our gap coverage estimate is one-fourth. According to TRADOC pamphlet 525-66 (U.S. Army, 2005b), FOC 11 has nine sub-requirements, and ATO#12 contributes to three of these, "improved sustainability," "enhancements in readiness, reliability, maintainability, and commonality for sustained operational tempo," and "global casualty care management." Accordingly, our estimate of $CV_{12,11}$ is

$$CV_{12,11} = 1/3 \times 1/4 \times 1/3 = 0.0278.$$

ATO#13: Intravenous Drug to Treat Severe and Complicated Malaria Caused by Multidrug-Resistant Malaria

The objective of this ATO is to develop a new intravenous drug to treat severe and complicated malaria caused by multidrug-resistant malaria parasites. It contributes to meeting capability requirements in FOC 7, personnel protection; and FOC 11, maneuver sustainment.

Since this drug could be used anywhere, according to Figure 3.2 ATO#13 applies to all three situations, off the battlefield, on the way to the battlefield, and on the battlefield. Therefore, there is no multiplier. The following describes our estimates of its CV contributions to each of the FOCs listed above.

FOC 7: Personnel Protection

Of the two categories in Table 3.1, ATO#13 applies only to "disease," so our gap coverage estimate is one-half. Moreover, malaria is only one of the five categories for disease defined in the notes of Table 3.1, so we multiply by an additional factor of one-fifth. There are no sub-requirements for this FOC, but we introduce an additional factor of one-half to provide redundancy for protection of the warfighter. Thus, our estimate of the CV contribution of ATO#13 to FOC 7 ($CV_{13,7}$) is

$$CV_{13,7} = 1/2 \times 1/5 \times 1/2 = 0.0500.$$

FOC 11: Maneuver Sustainment

Of the four categories in Table 3.1, ATO#13 applies only to "forces," so our gap coverage estimate is one-fourth. According to TRADOC pamphlet 525-66 (U.S. Army, 2005b), FOC 11 has nine sub-requirements, and ATO#13 contributes to three of these, "improved sustainability," "enhancements in readiness, reliability, maintainability, and commonality for sustained operational tempo," and "global casualty care management." Accordingly, our estimate of $CV_{13,11}$ is

$$CV_{13,11} = 1/4 \times 1/3 = 0.0833.$$

ATO#14: DNA Vaccines to Prevent Hemorrhagic Fevers

The objective of this ATO is to develop candidate vaccines to prevent hemorrhagic fevers such as hantavirus-associated hemorrhagic fever with renal syndrome and dengue virus-associated dengue fever and dengue hemorrhagic fever. It contributes to meeting capability requirements in FOC 7, personnel protection; and FOC 11, maneuver sustainment.

Since this vaccine could be used anywhere, according to Figure 3.2 ATO#13 applies to all three situations, off the battlefield, on the way to the battlefield, and on the battlefield. Therefore, there is no multiplier. The following describes our estimates of its CV contributions to each of the FOCs listed above.

FOC 7: Personnel Protection

Of the two categories in Table 3.1, ATO#14 applies only to "disease," so our gap coverage estimate is one-half. Moreover, infectious disease is only one of the five categories for disease defined in the notes of Table 3.1, so we multiply by an additional factor of one-fifth. There are no sub-requirements for this FOC, but we introduce an additional factor of one-half to provide redundancy for protection of the warfighter. Thus, our estimate of the CV contribution of ATO#14 to FOC 7 ($CV_{14,7}$) is

$$CV_{14,7} = 1/2 \times 1/5 \times 1/2 = 0.0500.$$

FOC 11: Maneuver Sustainment

Of the four categories in Table 3.1, ATO#14 applies only to "forces," so our gap coverage estimate is one-fourth. According to TRADOC pamphlet 525-66 (U.S. Army, 2005b), FOC 11 has nine sub-requirements, and ATO#14 contributes to two of these, "improved sustainability" and "enhancements in readiness, reliability, maintainability, and commonality for sustained operational tempo." Accordingly, our estimate of $CV_{14,11}$ is

$$CV_{14,11} = 1/4 \times 2/9 = 0.0556.$$

ATO#15: Fatigue Management Capability for Sustained Readiness and Performance

The objective of this ATO is to develop a predictive, quantitative fatigue management capability for mission planning, warrior performance assessment, and status reporting. It contributes to meeting capability requirements in FOC 7, personnel protection; and FOC 11, maneuver sustainment.

Since this capability applies to everyone, according to Figure 3.2 ATO#15 applies to all three situations: off the battlefield, on the way to the battlefield, and on the battlefield. Therefore, there is no multiplier. The following describes our estimates of its CV contributions to each of the FOCs listed above.

FOC 7 Personnel Protection

Of the two categories in Table 3.1, ATO#15 applies only to "injury," so our gap coverage estimate is one-half. Moreover, of the four categories for injury defined in the notes of Table 3.1, ATO#15 applies to all except "B/NLOS fire," so we multiply by an additional factor of three-fourths. There are no sub-requirements for this FOC, but we introduce an additional factor of one-half to provide redundancy for protection of the warfighter. Thus, our estimate of the CV contribution of ATO#15 to FOC 7 ($CV_{15,7}$) is

$$CV_{15,7} = 1/2 \times 3/4 \times 1/2 = 0.188.$$

FOC 11: Maneuver Sustainment

Of the four categories in Table 3.1, ATO#15 applies only to "forces," so our gap coverage estimate is one-fourth. According to TRADOC pamphlet 525-66 (U.S. Army, 2005b), FOC 11 has nine sub-requirements, and ATO#15 contributes to four of these, "improved sustainability," "enhancements in readiness, reliability, maintainability, and commonality for sustained operational tempo," "global force health and fitness," and "global casualty prevention." Accordingly, our estimate of $CV_{15,11}$ is

$$CV_{15,11} = 1/4 \times 4/9 = 0.111.$$

ATO#16: Physical Training Interventions to Enhance Military Task Performance and Reduce Musculoskeletal Injuries

The objective of this ATO is to provide new biomechanics- and physiology-based physical training and monitoring methods for quickly achieving the highest possible level of physical readiness while minimizing associated injury risk. It contributes to meeting capability requirements in FOC 7, personnel protection; and FOC 11, maneuver sustainment.

Since this training could be given to anyone, according to Figure 3.2 ATO#16 applies to all three situations, off the battlefield, on the way to the battlefield, and on the battlefield. Therefore, there is no multiplier. The following describes our estimates of its CV contributions to each of the FOCs listed above.

FOC 7: Personnel Protection

Of the two categories in Table 3.1, ATO#16 applies only to "injury," so our gap coverage estimate is one-half. Moreover, of the four categories for injury defined in the notes of Table 3.1, ATO#16 applies only to "accident," so we multiply by an additional factor of one-fourth. There are no sub-requirements for this FOC, but we introduce an additional factor of one-half to provide redundancy for protection of the warfighter. Thus, our estimate of the CV contribution of ATO#16 to FOC 7 ($CV_{16,7}$) is

$$CV_{16,7} = 1/2 \times 1/4 \times 1/2 = 0.0625.$$

FOC 11: Maneuver Sustainment

Of the four categories in Table 3.1, ATO#16 applies only to "forces," so our gap coverage estimate is one-fourth. According to TRADOC pamphlet 525-66 (U.S. Army, 2005b), FOC 11 has nine sub-requirements, and ATO#16 contributes to four of these, "improved sustainability," "enhancements in readiness, reliability, maintainability, and commonality for sustained operational tempo," "global force health and fitness," and "global casualty prevention." Accordingly, our estimate of $CV_{16,11}$ is

$$CV_{16,11} = 1/4 \times 4/9 = 0.111.$$

ATO#17: High-Altitude Airship

The objective of this ATO is to demonstrate the technical feasibility and military utility of an unmanned, untethered lighter-than-air airship carrying a multiple-mission payload that can fly at a 60,000 foot altitude above mean sea level. It contributes to meeting capability requirements in the following FOCs: FOC 2, battlespace awareness; FOC 3, mounted/dismounted maneuver; and FOC 6, maneuver support.

This multiple-mission ATO, according to Figure 3.2, applies to all three situations, off the battlefield, on the way to the battlefield, and on the battlefield. Therefore, there is no multiplier. The following describes our estimates of its CV contributions to each of the FOCs listed above.

FOC 2: Battlespace Awareness

Of the four categories in Table 3.1, ATO#17 applies to all, so there is no multiplier. According to TRADOC pamphlet 525-66 (U.S. Army, 2005b), FOC 2 has six sub-requirements, and ATO#17 contributes to only one of these, "ability to observe and collect information worldwide." Therefore, we multiply our gap coverage estimate by one-six. Thus, our estimate of the CV contribution of ATO#17 to FOC 2 ($CV_{17,2}$) is

$$CV_{17,2} = 1/6 = 0.167.$$

FOC 3: Mounted-Dismounted Maneuver

Of the four categories in Table 3.1, ATO#17 applies to "forces" and "weapons," so our gap coverage estimate is one-half. According to TRADOC pamphlet 525-66 (U.S. Army, 2005b), FOC 3 has ten sub-requirements, and ATO#17 contributes to two of these, "unsurpassed battlespace awareness" and "operations in urban and complex terrain." Therefore, we multiply our gap coverage estimate by one-fifth. Thus, our estimate of the CV contribution of ATO#17 to FOC 3 ($CV_{17,3}$) is

$$CV_{17,3} = 1/2 \times 1/5 = 0.100.$$

FOC 6: Maneuver Support

Of the four categories in Table 3.1, ATO#17 applies to "forces" and "weapons," so our gap coverage estimate is one-half. According to TRADOC pamphlet 525-66 (U.S. Army, 2005b), FOC 6 has seven sub-requirements, and ATO#17 contributes to only one of these, "understand battlespace environment." Accordingly, our estimate of $CV_{17,6}$ is

$$CV_{17,6} = 1/2 \times 1/7 = 0.0714.$$

ATO#18: Vaccine for Prevention of Military HIV Infection

The objective of this ATO is to transition to development a new-generation combination vaccine for the prevention of HIV infections in military forces. It contributes to meeting capability requirements in FOC 7, personnel protection; and FOC 11, maneuver sustainment.

Since this vaccine could be used anywhere, according to Figure 3.2 ATO#18 applies to all three situations, off the battlefield, on the way to the battlefield, and on the battlefield. Therefore, there is no multiplier. The following describes our estimates of its CV contributions to each of the FOCs listed above.

FOC 7: Personnel Protection

Of the two categories in Table 3.1, ATO#18 applies only to "disease," so our gap coverage estimate is one-half. Moreover, HIV/AIDS is only one of the five categories for disease defined in the notes of Table 3.1, so we multiply by an additional factor of one-fifth. There are no sub-requirements for this FOC, but we introduce an additional factor of one-half to provide redundancy for protection of the warfighter. Thus, our estimate of the CV contribution of ATO#18 to FOC 7 ($CV_{18,7}$) is

$$CV_{18,7} = 1/2 \times 1/5 \times 1/2 = 0.0500.$$

FOC 11: Maneuver Sustainment

Of the four categories in Table 3.1, ATO#18 applies only to "forces," so our gap coverage estimate is one-fourth. According to TRADOC pamphlet 525-66 (U.S. Army, 2005b), FOC 11 has nine sub-requirements, and ATO#18 contributes to two of these, "improved sustainability" and "enhancements in readiness, reliability, maintainability, and commonality for sustained operational tempo." Accordingly, our estimate of $CV_{18,11}$ is

$$CV_{18,11} = 1/4 \times 2/9 = 0.0556.$$

ATO#19: Joint Rapid Airfield Construction

The objective of this ATO is to develop materials and techniques for rapidly upgrading existing or constructing contingency airfields in-theater with a low logistics footprint. It contributes to meeting capability requirements in FOC 10, strategic responsiveness and deployability; and FOC 11, maneuver sustainment.

Since ATO#19 provides upgrades to airfields anywhere in-theater, according to Figure 3.2 it applies to all three situations, off the battlefield, on the way to the battlefield, and on the battlefield. Therefore, there is no multiplier. The following describes our estimates of its CV contributions to each of the FOCs listed above.

FOC 10: Strategic Responsiveness and Deployability

Of the three categories in Table 3.1, ATO#19 provides "transport" and "delivery," so our gap coverage estimate is two-thirds. According to TRADOC pamphlet 525-66 (U.S. Army, 2005b), FOC 10 has two sub-requirements, and ATO#19 applies only to one, "theater access enablers," so we multiply our gap coverage estimate by one-half. Accordingly, our estimate of $CV_{19,10}$ is

$$CV_{19,10} = 2/3 \times 1/2 = 0.333.$$

FOC 11: Maneuver Sustainment

Of the four categories in Table 3.1, ATO#19 applies to all, so there is no multiplier. According to TRADOC pamphlet 525-66 (U.S. Army, 2005b), FOC 11 has nine sub-requirements, and ATO#19 contributes to two of these, "improved sustainability" and "global precision delivery enhancements." Accordingly, our estimate of $CV_{19,11}$ is

$$CV_{19,11} = 2/9 = 0.222.$$

ATO#20: Dynamic Mission Readiness Training for C4ISR

The objective of this ATO is to ensure warfighter readiness through improved training and rehearsal capabilities, and to improve end-to-end interoperable communications through C2ISR process improvements. It contributes to meeting capability requirements in the following FOCs: FOC 1, battle command; FOC 3, mounted/dismounted maneuver; FOC 11, maneuver sustainment; and FOC 12, training, leadership, and education.

Because mission readiness training applies to all, according to Figure 3.2 ATO#20 applies to all three situations, off the battlefield, on the way to the battlefield, and on the battlefield. Therefore, there is no multiplier. The following describes our estimates of its CV contributions to each of the FOCs listed above.

FOC 1: Battle Command

Of the four categories in Table 3.1, ATO#20 applies to all, so there is no multiplier. According to TRADOC pamphlet 525-66 (U.S. Army, 2005b), FOC 1 has six sub-requirements, and ATO#20 contributes only to one of these, "decision planning and support capabilities." Therefore, we multiply our gap coverage estimate by one-sixth. Thus, our estimate of the CV contribution of ATO#20 to FOC 1 ($CV_{20,1}$) is

$$CV_{20,1} = 1/6 = 0.167.$$

FOC 3: Mounted-Dismounted Maneuver

Of the four categories in Table 3.1, ATO#20 applies only to "forces," so our gap coverage estimate is one-fourth. According to TRADOC pamphlet 525-66 (U.S. Army, 2005b), FOC 3 has ten sub-requirements, and ATO#20 contributes only to one of these, "quality, realistic, accessible training." Therefore, we multiply our gap coverage estimate by one-tenth. Thus, our estimate of the CV contribution of ATO#20 to FOC 3 ($CV_{20,3}$) is

$$CV_{20,3} = 1/4 \times 1/10 = 0.025.$$

FOC 11: Maneuver Sustainment

Of the four categories in Table 3.1, ATO#20 applies only to "forces," so our gap coverage estimate is one-fourth. According to TRADOC pamphlet 525-66 (U.S. Army, 2005b), FOC 11 has nine sub-requirements, and ATO#20 contributes only to one of these, "enhancements in readiness, reliability, maintainability, and commonality for sustained operational tempo." Accordingly, our estimate of $CV_{20,11}$ is

$$CV_{20,11} = 1/4 \times 1/9 = 0.0278.$$

FOC 12: Training, Leadership, and Education

Of the four categories in Table 3.1, ATO#20 applies to all, so there is no multiplier. According to TRADOC pamphlet 525-66 (U.S. Army, 2005b), FOC 12 has eight sub-requirements, and ATO#20 contributes to five of these, "accessible training," "realistic training," "responsive training development," "training for JIIM," and "managing unit performance." Accordingly, our estimate of $CV_{20,12}$ is

$$CV_{20,12} = 5/8 = 0.625.$$

ATO#21: Future Force Warrior ATD

The objective of this ATO is to develop and demonstrate revolutionary warfighting capabilities for soldiers and the small combat unit operating in the Future Force Unit of Action (UoA), and integrating the soldier within the UoA network for enhanced connectivity, synchronization, and application of force. It contributes to meeting capability requirements in the following FOCs: FOC 1, battle command; FOC 2, battlespace awareness; FOC 3, mounted/dismounted maneuver; FOC 5, LOS/BLOS/NLOS lethality; FOC 6, maneuver support; FOC 7, personnel protection; FOC 11, maneuver sustainment; FOC 12 training, leadership, and education; and FOC 13, human engineering.

Since ATO#21 is developing warfighting capabilities for soldiers and the small combat unit, according to Figure 3.2 it applies only to situation (3), on the battlefield. Therefore, we multiply its gap coverage estimates by one-third. The following describes our estimates of its CV contributions to each of the FOCs listed above.

FOC 1: Battle Command

Of the four categories in Table 3.1, ATO#21 applies to all, so there is no multiplier. According to TRADOC pamphlet 525-66 (U.S. Army, 2005b), FOC 1 has six sub-requirements, and ATO#21 contributes only to one of these, "networked force optimized for mobile operations." Therefore, we multiply our gap coverage estimate by one-sixth. Thus, our estimate of the CV contribution of ATO#21 to FOC 1 ($CV_{21,1}$) is

$$CV_{21,1} = 1/3 \times 1/6 = 0.0556.$$

FOC 2: Battlespace Awareness

Of the four categories in Table 3.1, ATO#21 applies to "LOS fire" and "enemy force location." Therefore, our gap coverage estimate is one-half. According to TRADOC pamphlet 525-66 (U.S. Army, 2005b), FOC 2 has six sub-requirements, and ATO#21 contributes only to one of these, "C2 of battlespace awareness assets." Therefore, we multiply our gap coverage estimate by one-six. Thus, our estimate of the CV contribution of ATO#21 to FOC 2 ($CV_{21,2}$) is

$$CV_{21,2} = 1/3 \times 1/2 \times 1/6 = 0.0278.$$

FOC 3: Mounted-Dismounted Maneuver

Of the four categories in Table 3.1, ATO#21 applies to "forces" and "weapons," so our gap coverage estimate is one-half. According to TRADOC pamphlet 525-66 (U.S. Army, 2005b), FOC 3 has ten sub-requirements, and ATO#21 contributes to five of these, "effective battle command," "unsurpassed battlespace awareness," "dependable and accurate LOS/BLOS/NLOS lethality," "sustainment with minimal load and logistics footprint," and "human engineering for improved soldier-system interface." Therefore, we multiply our gap coverage estimate by one-half. Thus, our estimate of the CV contribution of ATO#21 to FOC 3 ($CV_{21,3}$) is

$$CV_{21,3} = 1/3 \times 1/2 \times 1/2 = 0.0833.$$

FOC 5: LOS/BLOS/NLOS Lethality

Of the three categories in Table 3.1, ATO#21 applies only to "LOS lethality," so our gap coverage estimate is one-third. According to TRADOC pamphlet 525-66 (U.S. Army, 2005b), FOC 5 has two sub-requirements, and ATO#21 contributes to one of these, "LOS/BLOS lethality via precision, networked, responsive fires . . ." Therefore, we multiply our gap coverage estimate by one-half. Thus, our estimate of the CV contribution of ATO#21 to FOC 5 ($CV_{21,5}$) is

$$CV_{21,5} = 1/3 \times 1/3 \times 1/2 = 0.0555.$$

FOC 6: Maneuver Support

Of the four categories in Table 3.1, ATO#21 applies to "forces" and "weapons," so our gap coverage estimate is one-half. According to TRADOC pamphlet 525-66 (U.S. Army, 2005b), FOC 6 has seven sub-requirements, and ATO#21 contributes only to one of these, "understand battlespace environment." Accordingly, our estimate of $CV_{21,6}$ is

$$CV_{21,6} = 1/3 \times 1/2 \times 1/7 = 0.0238.$$

FOC 7: Personnel Protection

Of the two categories in Table 3.1, ATO#21 applies only to "injury," so our gap coverage estimate is one-half. Moreover, of the four categories of "injury" in the notes of Table 3.1, ATO#21 applies only to "LOS fire," so we multiply by an additional one-fourth. There are no sub-requirements for this FOC, but we introduce an additional factor of one-half to provide redundancy for protection of the warfighter. Accordingly, our estimate of $CV_{21,7}$ is

$$CV_{21,7} = 1/3 \times 1/2 \times 1/4 \times 1/2 = 0.0208.$$

FOC 11: Maneuver Sustainment

Of the four categories in Table 3.1, ATO#21 applies to "forces" and "weapons," so our gap coverage estimate is one-half. According to TRADOC pamphlet 525-66 (U.S. Army, 2005b), FOC 11 has nine sub-requirements, and ATO#21 contributes to two of these, "enhancements in readiness, reliability, maintainability, and commonality for sustained operational tempo," and "improved soldier support." Accordingly, our estimate of $CV_{21,11}$ is

$$CV_{21,11} = 1/3 \times 1/2 \times 2/9 = 0.0370.$$

FOC 12: Training, Leadership, and Education

Of the four categories in Table 3.1, ATO#21 applies to "forces" and "equipment," so our gap coverage estimate is one-half. According to TRADOC pamphlet 525-66 (U.S. Army, 2005b), FOC 12 has eight sub-requirements, and ATO#21 contributes only to one of these, "training for JIIM." Accordingly, our estimate of $CV_{21,12}$ is

$$CV_{21,12} = 1/3 \times 1/2 \times 1/8 = 0.0208.$$

FOC 13: Human Engineering

Of the three categories in Table 3.1, ATO#21 applies to all, so there is no multiplier. According to TRADOC pamphlet 525-66 (U.S. Army, 2005b), FOC 13 has four sub-requirements and ATO#21 contributes only to one of these, "reduce soldier dismounted movement approach load." However, ATO#21 contributes only to one of the three categories of soldier load defined in the notes of Table 3.1. Therefore, we multiply our gap coverage estimate by one-fourth and also by one-third. Thus, our estimate of the CV contribution of ATO#21 to FOC 13 ($CV_{21,13}$) is

$$CV_{21,13} = 1/3 \times 1/4 \times 1/3 = 0.0278.$$

ATO#22: Leader Adaptability

The objective of this ATO is to develop adaptable leaders through a combination of learning methods and training techniques, including computer-based interactive environments to mimic real-world experience. It contributes to meeting capability requirements in the following FOCs: FOC 1, battle command; FOC 11, maneuver sustainment; and FOC 12 training, leadership, and education.

Because leader adaptability applies to all, according to Figure 3.2 ATO#22 applies to all three situations, off the battlefield, on the way to the battlefield, and on the battlefield. Therefore, there is no multiplier. The following describes our estimates of its CV contributions to each of the FOCs listed above.

FOC 1: Battle Command

Of the four categories in Table 3.1, ATO#22 applies to "computers" and "command," so our gap coverage estimate is one-half. According to TRADOC pamphlet 525-66 (U.S. Army, 2005b), FOC 1 has six sub-requirements, and ATO#22 contributes only to one of these, "decision planning and support." Therefore, we multiply our gap coverage estimate by one-sixth. Thus, our estimate of the CV contribution of ATO#22 to FOC 1 ($CV_{22,1}$) is

$$CV_{22,1} = 1/2 \times 1/6 = 0.0833.$$

FOC 11: Maneuver Sustainment

Of the four categories in Table 3.1, ATO#22 applies to "forces" only, so our gap coverage estimate is one-fourth. According to TRADOC pamphlet 525-66 (U.S. Army, 2005b), FOC 11 has nine sub-requirements, and ATO#22 contributes only to one of these, "enhancements in readiness, reliability, maintainability, and commonality for sustained operational tempo." Accordingly, our estimate of $CV_{22,11}$ is

$$CV_{22,11} = 1/4 \times 1/9 = 0.0278.$$

FOC 12: Training, Leadership, and Education

Of the four categories in Table 3.1, ATO#22 applies to "forces" and "environment," so our gap coverage estimate is one-half. According to TRADOC pamphlet 525-66 (U.S. Army, 2005b), FOC 12 has eight sub-requirements, and ATO#22 contributes to three of these, "leadership training and education," "realistic training," and "training for JIIM." Accordingly, our estimate of $CV_{22,12}$ is

$$CV_{22,12} = 1/2 \times 3/8 = 0.188.$$

ATO#23: Advanced Antennas

The objective of this ATO is to address the substantial challenges in antenna performance (e.g., bandwidth, gain, mobility, signature) that must be overcome to achieve the full spectrum of communication effectiveness required for the highly mobile, mission-tailored, and combat-ready Future Force. The ATO contributes to meeting capability requirements in the following FOCs: FOC 1, battle command; FOC 2, battlespace awareness; FOC 3, mounted/dismounted maneuver; FOC 4, air maneuver; FOC 7, personnel protection; FOC 8, asset protection; and FOC 13, human engineering.

Because improved antenna performance applies to a wide variety of situations, according to Figure 3.2 ATO#23 applies to all three situations, off the battlefield, on the way to the battlefield, and on the battlefield. Therefore, there is no multiplier. The following describes our estimates of its CV contributions to each of the FOCs listed above.

FOC 1: Battle Command

Of the four categories in Table 3.1, ATO#23 applies only to "communications," so our gap coverage estimate is one-fourth. According to TRADOC pamphlet 525-66 (U.S. Army, 2005b), FOC 1 has six sub-requirements, and ATO#23 contributes to two of these, "layered, integrated C2 for joint, multinational, interagency operations on the move" and "networked force optimized for mobile operations." Therefore, we multiply our gap coverage estimate by one-third. Thus, our estimate of the CV contribution of ATO#23 to FOC 1 ($CV_{23,1}$) is

$$CV_{23,1} = 1/4 \times 1/3 = 0.0833.$$

FOC 2: Battlespace Awareness

Of the four categories in Table 3.1, ATO#23 applies to "LOS fire" and "enemy force location." Therefore, our gap coverage estimate is one-half. According to TRADOC pamphlet 525-66 (U.S. Army, 2005b), FOC 2 has six sub-requirements, and ATO#23 contributes to two of these, "C2 of battlespace awareness assets" and "ability to observe and collect information worldwide." Therefore, we multiply our gap coverage estimate by one-third. Thus, our estimate of the CV contribution of ATO#23 to FOC 2 ($CV_{23,2}$) is

$$CV_{23,2} = 1/2 \times 1/3 = 0.167.$$

FOC 3: Mounted-Dismounted Maneuver

Of the four categories in Table 3.1, ATO#23 applies to "equipment" and "mobility," so our gap coverage estimate is one-half. According to TRADOC pamphlet 525-66 (U.S. Army, 2005b), FOC 3 has ten sub-requirements, and ATO#23 contributes to three of these, "effective battle command," "unsurpassed battlespace awareness," and "sustainment with minimal load and logistics footprint." Therefore, we multiply our gap coverage estimate by three-tenths. Thus, our estimate of the CV contribution of ATO#23 to FOC 3 ($CV_{23,3}$) is

$$CV_{23,3} = 1/2 \times 3/10 = 0.150.$$

FOC 4: Air Maneuver

Of the four categories in Table 3.1, ATO#23 applies to "equipment" and "mobility," so our gap coverage estimate is one-half. According to TRADOC pamphlet 525-66 (U.S. Army, 2005b), FOC 4 has five sub-requirements, and ATO#24 contributes only to one of these, "assured and timely connectivity with the supported force." Therefore, we multiply our gap coverage estimate by one-fifth. Thus, our estimate of the CV contribution of ATO#23 to FOC 4 ($CV_{23,4}$) is

$$CV_{23,4} = 1/2 \times 1/5 = 0.100.$$

FOC 7: Personnel Protection

Of the two categories in Table 3.1, ATO#23 applies only to "injury," so our gap coverage estimate is one-half. Moreover, of the four categories of "injury" in the notes of Table 3.1, ATO#23 applies only to "LOS fire," so we multiply by an additional one-fourth. There are no sub-requirements for this FOC, but we introduce an additional factor of one-half to provide redundancy for "personnel protection." Accordingly, our estimate of $CV_{23,7}$ is

$$CV_{23,7} = 1/2 \times 1/4 \times 1/2 = 0.0625.$$

FOC 8: Asset Protection

Of the four categories in Table 3.1, ATO#23 applies only to "LOS fire," so our gap coverage estimate is one-fourth. There are no sub-requirements for this FOC, but we introduce an additional factor of one-half to provide redundancy for "protection of warfighting assets." Accordingly, our estimate of $CV_{23,8}$ is

$$CV_{23,8} = 1/4 \times 1/2 = 0.125.$$

FOC 13: Human Engineering

Of the three categories in Table 3.1, ATO#23 applies to engineering of the "systems" and "human-system interfaces," so our gap coverage estimate is two-thirds. According to TRADOC pamphlet 525-66 (U.S. Army, 2005b), FOC 13 has four sub-requirements and ATO#23 contributes only to one of these, "reduce soldier dismounted movement approach load." However, ATO#23 contributes only to one of the three categories of soldier load defined in the notes of Table 3.1. Therefore, we multiply our gap coverage estimate by one-fourth and also by one-third. Thus, our estimate of the CV contribution of ATO#23 to FOC 13 ($CV_{23,13}$) is

$$CV_{23,13} = 2/3 \times 1/4 \times 1/3 = 0.0556.$$

ATO#24: Autonomous Distributed Sensors

The objective of this ATO is to demonstrate a capability to detect, track, classify, and report battlefield and undersea threats with distributed unmanned sensors using multispectral acoustic, magnetic, and seismic emission from the threat. It contributes to meeting capability requirements in the following FOCs: FOC 2, battlespace awareness; FOC 6, maneuver support; FOC 7, personnel protection; and FOC 8, asset protection.

Because these sensors can be distributed anywhere, according to Figure 3.2 ATO#24 applies to all three situations, off the battlefield, on the way to the battlefield, and on the battlefield. Therefore, there is no multiplier. The following describes our estimates of its CV contributions to each of the FOCs listed above.

FOC 2: Battlespace Awareness

Of the four categories in Table 3.1, ATO#24 applies to all, so there is no multiplier. According to TRADOC pamphlet 525-66 (U.S. Army, 2005b), FOC 2 has six sub-requirements, and ATO#24 contributes to two of these, "ability of observe and collect information worldwide" and "fusion of information." Therefore, we multiply our gap coverage estimate by one-third. Thus, our estimate of the CV contribution of ATO#24 to FOC 2 ($CV_{24,2}$) is

$$CV_{24,2} = 1/3 = 0.333.$$

FOC 6: Maneuver Support

Of the four categories in Table 3.1, ATO#24 applies only to "forces," so our gap coverage estimate is one-fourth. According to TRADOC pamphlet 525-66 (U.S. Army, 2005b), FOC 6 has seven sub-requirements, and ATO#24 contributes to two of these, "detect and neutralize environmental hazards" and "understand battlespace environment." Accordingly, our estimate of $CV_{24,6}$ is

$$CV_{24,6} = 1/4 \times 2/7 = 0.0714.$$

FOC 7: Personnel Protection

Of the two categories in Table 3.1, ATO#24 applies only to "injury," so our gap coverage estimate is one-half. Moreover, of the four categories of "injury" in the notes of Table 3.1, ATO#24 applies to "LOS fire," "NLOS fire," and "hazards," so we multiply by an additional three-fourths. There are no sub-requirements for this FOC, but we introduce an additional factor of one-half to provide redundancy for "personnel protection." Accordingly, our estimate of $CV_{24,7}$ is

$$CV_{24,7} = 1/2 \times 3/4 \times 1/2 = 0.188.$$

FOC 8: Asset Protection

Of the four categories in Table 3.1, ATO#24 applies to "LOS fire," "NLOS fire," and "hazards," so our gap coverage estimate is three-fourths. There are no sub-requirements for this FOC, but we introduce an additional factor of one-half to provide redundancy for "protection of warfighting assets." Accordingly, our estimate of $CV_{24,8}$ is

$$CV_{24,8} = 3/4 \times 1/2 = 0.375.$$

ATO#25: Tactical Wireless Network Assurance

The objective of this ATO is to develop information assurance improvements to defeat sophisticated information warfare threats launched against the emerging wireless tactical networks that must provide critical mobile communications for the transformational force. It contributes to meeting capability requirements in the following FOCs: FOC 1, battle command; FOC 4, air maneuver; and FOC 9, information protection.

Because "information assurance" applies to everyone, according to Figure 3.2 ATO#25 applies to all three situations, off the battlefield, on the way to the battlefield, and on the battlefield. Therefore, there is no multiplier. The following describes our estimates of its CV contributions to each of the FOCs listed above.

FOC 1: Battle Command

Of the four categories in Table 3.1, ATO#25 applies only to "communications," so our gap coverage estimate is one-fourth. According to TRADOC pamphlet 525-66 (U.S. Army, 2005b), FOC 1 has six sub-requirements, and ATO#25 contributes only to one of these, "information protection and rapid restoration of information systems." Therefore, we multiply our gap coverage estimate by one-sixth. Thus, our estimate of the CV contribution of ATO#25 to FOC 1 ($CV_{25,1}$) is

$$CV_{25,1} = 1/4 \times 1/6 = 0.0417.$$

FOC 4: Air Maneuver

Of the four categories in Table 3.1, ATO#25 applies only to "equipment," so our gap coverage estimate is one-fourth. According to TRADOC pamphlet 525-66 (U.S. Army, 2005b), FOC 4 has five sub-requirements, and ATO#25 contributes only to one of these, "assured and timely connectivity with the supported force." Therefore, we multiply our gap coverage estimate by one-fifth. Thus, our estimate of the CV contribution of ATO#25 to FOC 4 ($CV_{25,4}$) is

$$CV_{25,4} = 1/4 \times 1/5 = 0.0500.$$

FOC 9: Information Protection

Of the four categories in Table 3.1, ATO#25 applies to all, so there is no multiplier. There are no sub-requirements for this FOC, but we introduce an additional factor of one-half to provide redundancy for "protection of information." Accordingly, our estimate of $CV_{25,9}$ is

$$CV_{25,9} = 1/2 = 0.500.$$

ATO#26: Command, Control, and Communications On-the-Move Demonstration

The objective of this ATO is to develop and demonstrate the warfighting value of an integrated command, control, and communications (C3) on-the-move capability through a series of field demonstrations. It contributes to meeting capability requirements in the following FOCs: FOC 1, battle command; FOC 2, battlespace awareness; FOC 3, mounted/dismounted maneuver; FOC 5, LOS/BLOS/NLOS lethality; FOC 6, maneuver support; FOC 7, personnel protection; FOC 8, asset protection; and FOC 11, maneuver sustainment.

Because integrated C3 applies to a wide variety of situations, according to Figure 3.2 ATO#26 applies to all three situations, off the battlefield, on the way to the battlefield, and on the battlefield. Therefore, there is no multiplier. The following describes our estimates of its CV contributions to each of the FOCs listed above.

FOC 1: Battle Command

Of the four categories in Table 3.1, ATO#26 applies to all, so there is no multiplier. According to TRADOC pamphlet 525-66 (U.S. Army, 2005b), FOC 1 has six subrequirements, and ATO#26 contributes to four of these, "layered, integrated C2 for joint, multinational, interagency operations on the move," "networked force optimized for mobile operations," "decision planning and support capabilities," and "information operations integrated with information management and ISR." Therefore, we multiply our gap coverage estimate by two-thirds. Thus, our estimate of the CV contribution of ATO#26 to FOC 1 ($CV_{26,1}$) is

$$CV_{26,1} = 2/3 = 0.667.$$

FOC 2: Battlespace Awareness

Of the four categories in Table 3.1, ATO#26 applies to all, so there is no multiplier. According to TRADOC pamphlet 525-66 (U.S. Army, 2005b), FOC 2 has six subrequirements, and ATO#26 contributes to two of these, "C2 of battlespace awareness assets," and "fusion of information." Therefore, we multiply our gap coverage estimate by one-third. Thus, our estimate of the CV contribution of ATO#26 to FOC 2 ($CV_{26,2}$) is

$$CV_{26,2} = 1/3 = 0.333.$$

FOC 3: Mounted-Dismounted Maneuver

Of the four categories in Table 3.1, ATO#26 applies to "forces," "mobility," and "weapons," so our gap coverage estimate is three-fourths. According to TRADOC pamphlet 525-66 (U.S. Army, 2005b), FOC 3 has ten sub-requirements, and ATO#26 contributes to two of these, "effective battle command" and "unsurpassed battlespace awareness." Therefore, we multiply our gap coverage estimate by one-fifth. Thus, our estimate of the CV contribution of ATO#26 to FOC 3 ($CV_{26,3}$) is

$$CV_{26,3} = 3/4 \times 1/5 = 0.150.$$

FOC 5: LOS/BLOS/NLOS Lethality

Of the three categories in Table 3.1, ATO#26 applies to "LOS lethality" and "BLOS lethality," so our gap coverage estimate is two-thirds. According to TRADOC pamphlet 525-66 (U.S. Army, 2005b), FOC 5 has two sub-requirements, and ATO#26 contributes to one of these, "LOS/BLOS lethality via precision, networked, responsive fires . . ." Therefore, we multiply our gap coverage estimate by one-half. Thus, our estimate of the CV contribution of ATO#26 to FOC 5 ($CV_{26,5}$) is

$$CV_{26,5} = 2/3 \times 1/2 = 0.333.$$

FOC 6: Maneuver Support

Of the four categories in Table 3.1, ATO#26 applies to "forces," "mobility," and "weapons," so our gap coverage estimate is three-fourths. According to TRADOC pamphlet 525-66 (U.S. Army, 2005b), FOC 6 has seven sub-requirements, and ATO#26 contributes only to one of these, "understand battlespace environment." Accordingly, our estimate of $CV_{26,6}$ is

$$CV_{26,6} = 3/4 \times 1/7 = 0.107.$$

FOC 7: Personnel Protection

Of the two categories in Table 3.1, ATO#26 applies only to "injury," so our gap coverage estimate is one-half. Moreover, of the four categories of "injury" in the notes of Table 3.1, ATO#26 applies to "LOS fire" and "hazards," so we multiply by an additional one-half. There are no sub-requirements for this FOC, but we introduce an additional factor of one-half to provide redundancy for "personnel protection." Accordingly, our estimate of $CV_{26,7}$ is

$$CV_{26,7} = 1/2 \times 1/2 \times 1/2 = 0.125.$$

FOC 8: Asset Protection

Of the four categories in Table 3.1, ATO#26 applies to "LOS fire" and "hazards," so our gap coverage estimate is one-half. There are no sub-requirements for this FOC, but we introduce an additional factor of one-half to provide redundancy for protection of warfighting assets. Accordingly, our estimate of $CV_{26,8}$ is

$$CV_{26,8} = 1/2 \times 1/2 = 0.250.$$

FOC 11: Maneuver Sustainment

Of the four categories in Table 3.1, ATO#26 applies to "forces," "mobility," and "weapons," so our gap coverage estimate is three-fourths. According to TRADOC pamphlet 525-66 (U.S. Army, 2005b), FOC 11 has nine sub-requirements, and ATO#26 contributes only to one of these, "global casualty prevention." Accordingly, our estimate of $CV_{26,11}$ is

$$CV_{26,11} = 3/4 \times 1/9 = 0.0833.$$

ATO#27: Modeling Architecture for Technology, Research, and Experimentation

The objective of this ATO is to develop an interoperable simulation architecture and reference implementation using standardized component interfaces and processes that represent key characteristics of network-centric warfighting systems. It contributes to meeting capability requirements in the following FOCs: FOC 1, battle command; FOC 2, battlespace awareness; and FOC 11, maneuver sustainment.

Because this simulation architecture applies to a wide variety of situations, according to Figure 3.2 ATO#27 applies to all three situations, off the battlefield, on the way to the battlefield, and on the battlefield. Therefore, there is no multiplier. The following describes our estimates of its CV contributions to each of the FOCs listed above.

FOC 1: Battle Command

Of the four categories in Table 3.1, ATO#27 applies only to "computers," so our gap coverage estimate is one-fourth. According to TRADOC pamphlet 525-66 (U.S. Army, 2005b), FOC 1 has six sub-requirements, and ATO#27 contributes only to one of these, "decision planning and support capabilities." Therefore, we multiply our gap coverage estimate by one-sixth. Thus, our estimate of the CV contribution of ATO#27 to FOC 1 ($CV_{27,1}$) is

$$CV_{27,1} = 1/4 \times 1/6 = 0.0417.$$

FOC 2: Battlespace Awareness

Of the four categories in Table 3.1, ATO#27 applies to "LOS fire," "enemy force location," and "hazards," so our gap coverage estimate is three-fourths. According to TRADOC pamphlet 525-66 (U.S. Army, 2005b), FOC 2 has six sub-requirements, and ATO#27 contributes only to one of these, "ability to model, simulate, and forecast." Therefore, we multiply our gap coverage estimate by one-sixth. Thus, our estimate of the CV contribution of ATO#27 to FOC 2 ($CV_{27,2}$) is

$$CV_{27,2} = 3/4 \times 1/6 = 0.125.$$

FOC 11: Maneuver Sustainment

Of the four categories in Table 3.1, ATO#27 applies only to "forces," so our gap coverage estimate is one-fourth. According to TRADOC pamphlet 525-66 (U.S. Army, 2005b), FOC 11 has nine sub-requirements, and ATO#27 contributes only to one of these, "enhancements in readiness, reliability, maintainability, and commonality for sustained operational tempo." Accordingly, our estimate of $CV_{27,11}$ is

$$CV_{27,11} = 1/4 \times 1/9 = 0.0278.$$

ATO#28: Radio-Enabling Technologies and Nextgen Applications—Joint ATD

The objective of this ATO is to improve radio communications by addressing critical capability challenges and shortfalls of the Joint Tactical Radio System clustered programs. It contributes to meeting capability requirements in the following FOCs: FOC 1, battle command; FOC 3, mounted/dismounted maneuver; FOC 4, air maneuver; and FOC 11, maneuver sustainment.

Because it improves radio communication for all, according to Figure 3.2 ATO#28 applies to all three situations, off the battlefield, on the way to the battlefield, and on the battlefield. Therefore, there is no multiplier. The following describes our estimates of its CV contributions to each of the FOCs listed above.

FOC 1: Battle Command

Of the four categories in Table 3.1, ATO#28 applies only to "communications," so our gap coverage estimate is one-fourth. According to TRADOC pamphlet 525-66 (U.S. Army, 2005b), FOC 1 has six sub-requirements, and ATO#28 contributes to two of these, "layered, integrated C2 for joint, multinational, interagency operations on the move" and "networked force optimized for mobile operations." Therefore, we multiply our gap coverage estimate by one-third. Thus, our estimate of the CV contribution of ATO#29 to FOC 1 ($CV_{28,1}$) is

$$CV_{28,1} = 1/4 \times 1/3 = 0.0833.$$

FOC 3: Mounted-Dismounted Maneuver

Of the four categories in Table 3.1, ATO#28 applies to "forces," so our gap coverage estimate is one-fourth. According to TRADOC pamphlet 525-66 (U.S. Army, 2005b), FOC 3 has ten sub-requirements, and ATO#28 contributes only to one of these, "effective battle command." Therefore, we multiply our gap coverage estimate by one-tenth. Thus, our estimate of the CV contribution of ATO#28 to FOC 3 ($CV_{28,3}$) is

$$CV_{28,3} = 1/4 \times 1/10 = 0.0250.$$

FOC 4: Air Maneuver

Of the four categories in Table 3.1, ATO#28 applies to "forces," so our gap coverage estimate is one-fourth. According to TRADOC pamphlet 525-66 (U.S. Army, 2005b), FOC 4 has five sub-requirements, and ATO#28 contributes only to one of these, "assured and timely connectivity with the supported force." Therefore, we multiply our gap coverage estimate by one-fifth. Thus, our estimate of the CV contribution of ATO#28 to FOC 4 ($CV_{28,4}$) is

$$CV_{28,4} = 1/4 \times 1/5 = 0.0500.$$

FOC 11: Maneuver Sustainment

Of the four categories in Table 3.1, ATO#28 applies only to "forces," so our gap coverage estimate is one-fourth. According to TRADOC pamphlet 525-66 (U.S. Army, 2005b), FOC 11 has nine sub-requirements, and ATO#28 contributes to two of these, "enhancements in readiness, reliability, maintainability, and commonality for sustained operational tempo" and "improved soldier support." Accordingly, our estimate of $CV_{28,11}$ is

$$CV_{28,11} = 1/4 \times 2/9 = 0.0556.$$

ATO#29: On-Route Detection of Mines/IEDs

The objective of this ATO is to integrate unmanned aerial vehicle standoff change detection and precision wideband ground-penetrating radar for high-confidence, high-operational-tempo detection of on-route antitank landmines and roadside improvised explosive devices (IEDs). It contributes to meeting capability requirements in the following FOCs: FOC 3, mounted/dismounted maneuver; FOC 6, maneuver support; FOC 7, personnel protection; FOC 8, asset protection; and FOC 11, maneuver sustainment.

Because it detects on-route, according to Figure 3.2 ATO#29 applies to situations (2) and (3), on the way to the battlefield and on the battlefield. Therefore, we multiply our gap coverage estimates by two-thirds. The following describes our estimates of its CV contributions to each of the FOCs listed above.

FOC 3 Mounted-Dismounted Maneuver
Of the four categories in Table 3.1, ATO#29 applies to "forces" and "mobility," so our gap coverage estimate is one-half. According to TRADOC pamphlet 525-66 (U.S. Army, 2005b), FOC 3 has ten sub-requirements, and ATO#29 contributes only to one of these, "operations in urban and complex terrain." Therefore, we multiply our gap coverage estimate by one-tenth. Thus, our estimate of the CV contribution of ATO#29 to FOC 3 ($CV_{29,3}$) is

$$CV_{29,3} = 2/3 \times 1/2 \times 1/10 = 0.0333.$$

FOC 6: Maneuver Support
Of the four categories in Table 3.1, ATO#29 applies to "forces" and "mobility," so our gap coverage estimate is one-half. According to TRADOC pamphlet 525-66 (U.S. Army, 2005b), FOC 6 has seven sub-requirements, and ATO#29 contributes to two of these, "provide assured mobility" and "detect and neutralize environmental hazards." Accordingly, our estimate of $CV_{29,6}$ is

$$CV_{29,6} = 2/3 \times 1/2 \times 2/7 = 0.0952.$$

FOC 7: Personnel Protection
Of the two categories in Table 3.1, ATO#29 applies only to "injury," so our gap coverage estimate is one-half. Moreover, of the four categories of "injury" in the notes of Table 3.1, ATO#29 applies only to "hazards," so we multiply by an additional one-fourth. There are no sub-requirements for this FOC, but we introduce an additional factor of one-half to provide redundancy for personnel protection. Accordingly, our estimate of $CV_{29,7}$ is

$$CV_{29,7} = 2/3 \times 1/2 \times 1/4 \times 1/2 = 0.0417.$$

FOC 8: Asset Protection

Of the four categories in Table 3.1, ATO#29 applies only to "hazards," so our gap coverage estimate is one-fourth. There are no sub-requirements for this FOC, but we introduce an additional factor of one-half to provide redundancy for protection of warfighting assets. Accordingly, our estimate of $CV_{29,8}$ is

$$CV_{29,8} = 2/3 \times 1/4 \times 1/2 = 0.0833.$$

FOC 11: Maneuver Sustainment

Of the four categories in Table 3.1, ATO#29 applies to "forces" and "mobility," so our gap coverage estimate is one-half. According to TRADOC pamphlet 525-66 (U.S. Army, 2005b), FOC 11 has nine sub-requirements, and ATO#29 contributes to two of these, "improved sustainability" and "global casualty prevention." Accordingly, our estimate of $CV_{29,11}$ is

$$CV_{29,11} = 2/3 \times 1/2 \times 2/9 = 0.0741.$$

Estimation of Marginal Unit Cost, Number of Units, and Marginal Operating and Maintenance Cost

For a description of our selected ATOs, which are also DTOs, we rely on the DTO data sheets. These data sheets are updated regularly on the Web by DDR&E, and the version we used is dated 2006, the latest available during our study. Because this source provides mostly technical data, we use other sources for cost, as indicated throughout this appendix, as well as additional technical data pertaining to our selected ATOs and systems to be derived from these ATOs.

Marginal unit cost (MUC) is the cost of producing one unit of an ATO-derived system, as opposed to continuing to produce the legacy system that is to be replaced by the new system. This cost can be positive if the system is more expensive than the legacy system, zero if its cost is the same as the cost of the legacy system, or negative if the new system is less costly than the legacy system that it replaces.

Costs associated with producing and maintaining weapon systems are considered in this estimation, so that changes in the cost of producing and maintaining physical items, as well as changes in manpower required to operate systems, are all taken into account. The RDT&E (S&T, SDD, and upgrade) costs associated with developing systems are not covered in this appendix. Instead, they are covered in Chapter Four and Appendix D.

If there is no existing system that serves the same purposes as the new system, the cost-over-legacy numbers simply indicate the costs of producing the new system. For example, no vaccine is currently effective against human immunodeficiency virus (HIV). Introducing an HIV vaccine into production generates a new cost over and above what the Army would incur if it simply did not produce the vaccine. As a result, the cost of producing the vaccine is considered the cost over legacy of the vaccine.[1]

For each new weapon system analyzed, units to be produced over a 20-year planning horizon can be estimated by equating the new system to an existing system serving similar functions and for which the number of units produced is known. An

---

[1] At the same time, while there is no cost subtraction, there is also no subtraction in the new vaccine's capability contribution in meeting the FOC requirements. In other words, we measure both cost and contribution in marginal terms over and above the legacy system to be replaced.

alternative method is to estimate directly the future demand of the new system. If a comparison system[2] is available, units to be procured are based on the units acquired for the comparison system over a 20-year period.[3] If no comparison system is available, a projection method is used to estimate units produced. This method is described in Appendix D.

The marginal procurement cost (MPC) is the difference between the procurement cost of the new systems and that of the legacy systems over a 20-year period. It is equal to marginal unit cost times the number of units procured.

The MPC can then be used to calculate MOMC. The operating and maintenance cost includes manning, operating, and maintaining all units during their operating life, when these units are procured over a 20-year period. To calculate MOMC, an O&M percentage is often used. The O&M cost of the new systems over a 20-year period is expressed as a percentage of these systems' total procurement cost. Mechanical and vehicular systems that are subject to use in outdoor environments are expected to have high operating and maintenance expenses totaling 115 percent of their unit cost, over their operating lifetime. Systems that are generally not handled manually but that require regular upkeep or maintenance are assigned a percentage of 80 percent of unit cost. Finally, systems that require only a minimum amount of storage, or software systems that require no manual maintenance, are assigned a percentage of 10 percent.[4]

Once legacy O&M cost is determined, a small fraction is added or subtracted from it based on the profile of the new weapon system. For example, Small Unit Operations produces a radio that is expected to have lower O&M cost than the legacy radios to be replaced. Assigning the old radio an O&M percentage (OMCLEG%) of 115 percent, the new radio is assigned a lower O&M percentage (OMCNEW%) of 105 percent. If no legacy system exists for the ATO, then this procedure does not apply. Instead, the ATO is simply assigned one of the O&M percentages outlined above and in Chapter Four (150, 115, 80, 45, or 10 percent).

The O&M cost for a system is determined by the nature of the weapon system in question. First, if the new system replaces another and is expected to have the same operating and maintenance costs, MOMC is simply set to zero.

Second, if the system is expected to add complexity above a legacy system, MOMC will be positive. MOMC is equal to the total acquisition cost of the new systems over the 20-year period times an O&M percentage for the new systems (OMCNEW%) minus

---

[2]   A comparison system is one that we consider to be similar to the ATO-derived system in terms of functions and number of units needed.

[3]   As the military is still acquiring additional units of many of the systems under comparison, an extrapolation scheme was used to estimate the total number of units to be acquired. This and other extrapolation schemes are discussed in Appendix D.

[4]   See the subsection "Marginal Operating and Maintenance Cost" in Chapter Four for the other two O&M cost levels: 45 and 150 percent.

total acquisition cost of the old systems times the old O&M percentage (OMCLEG%), if a legacy system exists. Otherwise, the second term is zero. This second calculation procedure is referred to as the "standard method" below.

A third method is best explained with an example. ATO #8, Power and Energy, introduces hybrid vehicles that would replace legacy vehicular systems. The hybrid system is more complex, but that complexity is not expected to lead to a significantly higher level of O&M. Instead, the DTO data sheet indicates that one purpose of the Power and Energy ATO is to reduce O&M costs when compared to legacy systems. This reduction in O&M cost is based on the total acquisition cost of the legacy systems. Therefore, in this case, the method to calculate MOMC is

$$(\text{OMCNEW\%} \times \text{NPLEG} \times \text{UCLEG}) - (\text{OMCLEG\%} \times \text{NPLEG} \times \text{UCLEG}),$$

where NPLEG is the number of legacy systems procurement over the 20-year period, and UCLEG is the unit cost of a legacy system.

Below we estimate the cost components for each ATO.

ATO#1 Small Unit Operations TD

Cost over Legacy (MUC)[5]

The radio unit produced under this ATO is expected to replace some capabilities of existing radios but with better performance. We compared it to SINCGARS in determining cost-over-legacy estimations. We expect that over a 20-year period, producing the radio will result in costs similar to those of the existing radios. As a result, cost over legacy or marginal unit cost in this case is zero.

O&M

The 2006 *Defense Technology Objectives* specifically mentions that this weapon is being developed in part to reduce maintenance cost of older and less reliable communications systems. We assigned an OMCLEG% of 115 percent to the legacy system, and considered the new system to have its OMCNEW% at a slightly lower percentage of 105 percent. O&M is calculated by the standard method and falls by $319 million.

Units to Be Procured

A total of 265,607 units are expected to be produced under this ATO. This is equal to the number of SINCGARS units produced over a 20-year period.

---

[5] We use "cost over legacy" and the marginal unit cost interchangeably; both mean the cost of procuring one unit of a new system minus that of a legacy system that the new system replaces.

ATO#2 Vaccines for the Prevention of Malaria

Cost over Legacy

Currently, there is no vaccine to prevent malaria, so the full cost of the vaccine is equal to cost over legacy. Like other vaccines under development in other ATOs, this is described as a DNA (Deoxyribonucleic Acid) vaccine, whose cost is likely to be higher than a regular vaccine. We assume a dose of DNA vaccine to be $5.

O&M

The O&M percentage associated with this system is 10 percent since maintenance mainly associated with storage is likely to be low. O&M is calculated by the standard method and results in a cost of $858 thousand.

Units to Be Procured

DNA vaccines are expected to be effective for nine months per injection, which should cover an average tour of duty (Touchette, 2004). The approximate number of troops currently in Afghanistan and Iraq that require vaccination is 170,000 ("Surviving a Tour of Duty," 2005). An assumption of four such contingencies over the next 20 years leads to 680,000 total doses. In any given year, 6,000 soldiers will be stationed elsewhere in the Middle East.[6] Another 97,000 will be stationed in Asia, and of these, 45,000 will be stationed outside of Japan and thus at risk of contracting malaria. Finally, around 800 troops a year will be stationed in at-risk areas in Africa. In total over 20 years, 1,716,000 vaccine doses will be needed.

ATO#3 Overwatch

Cost over Legacy

Like Overwatch, the existing ATFLIR sensor system utilizes an IR sensor in locating ground targets. However, ATFLIR is of longer range and affixed to fast-moving aircraft, rather than to the slower-moving helicopters and ground vehicles for which Overwatch is suited. One ATFLIR system procurement cost is $3.126 million (Nicholas and Rossi, 2007). The cost of an Overwatch system is likely to be half of that cost, or $1.56 million.

O&M

Overwatch is an electronics system used outdoors. An O&M percentage of 115 percent is assigned to the system, and the standard method is used in calculation. O&M is calculated by the standard method and is $7.2 billion.

---

[6]   The numbers of troops here and below in this paragraph come from *GlobalSecurity.org* as of November 24, 2007.

Units to Be Procured

In Chapter Four, we described how we estimated the number of units produced to be 4,000. Here, we discuss the estimate in more detail. Currently, there are 17,000 HMMWVs in Iraq (Shalal-Esa, 2007). We assume that one in ten will be equipped with an Overwatch system to provide attack warning to a convoy or during other missions (Vasquez, 2005). Thus, 1,700 HMMWVs used in a future contingency similar to Operation Iraqi Freedom (OIF) will be equipped with Overwatch systems.

Besides ground vehicles, the Overwatch system is slated to be used on unmanned aerial vehicles (UAVs), helicopters, watercraft, and aerostats (Christensen, 2006). Moreover, we expect the system to be used in other theaters. Together, we assume 20 percent more for use on other platforms and theaters. This brings the number of units to 2,000 with rounding.

We also studied the Boomerang system, which picks up sniper locations acoustically. It is currently installed on 200 vehicles in Iraq and Afghanistan—an order of magnitude lower than our estimate of 2,000. However, we expect that the Overwatch system is much more capable in detecting and locating both small-caliber and large-caliber munitions and will be in greater demand.

Considering that the Overwatch systems will be operating in an outdoor and combat environment, we expect that many will not last for 20 years and some will be destroyed by enemy fire. We assume that they will be replaced once over the 20-year period, bringing the total number to 4,000.

ATO#4: Unmanned Ground Mobility

Cost over Legacy

The leader-follower capability provided by the system is similar to the auto-flight capability found in UAVs. As a rough approximation, the Dragoneye UAV is a similar small, automated technology. Based on the description in the original DTO, the leader-follower capability initially developed under the Unmanned Ground Mobility system is not likely to be very complicated, although further refinements are planned for the system in the future.

Regardless of the eventual cost of the Unmanned Ground Mobility system, it must be low enough for installation in all convoy follower vehicles. The cost of one Dragoneye system—roughly one hundred thousand dollars—is used as the proxy for the eventual cost of the Unmanned Ground Mobility system (Nicholas and Rossi, 2007).

O&M

The system will be contained within military vehicles. As a result, O&M should be lower than for a system exposed to the elements, and an O&M percentage of 80 percent is assigned to the system. O&M is calculated by the standard method to be $320 million.

Units to Be Procured

An average convoy in Iraq consists of 30 vehicles (Vasquez, 2005). Twenty-nine of those vehicles will require the follower and self-navigating capability provided by Unmanned Ground Mobility. There are 20 to 30 convoys daily in Iraq (Michaels, 2007), suggesting a total of 25 convoys of 29 follower vehicles each, or 725 systems necessary in daily operations.

If there are five times as many vehicles ready to convoy as are required to each day, then 3,625 vehicles will be fitted with the system. However, not all convoys will take place on high-risk routes. If only half do, then the estimate is cut to 1,813 vehicles requiring the system. In turn, other conflicts will require some vehicles fitted with the system, but far fewer than necessary in a future OIF-like contingency. Rounding to 2,000 systems should account for these requirements.

Convoy vehicles are likely to last for ten years. Over the full 20-year planning horizon, twice as many systems will be necessary, resulting in a final estimate of 4,000 systems procured.

ATO#5 Warfighter-Systems Interaction

Cost over Legacy

The system may be used to replace AFATDS, while adding some features. One AFATDS unit costs $109,000 (Nicholas and Rossi, 2007). The Warfighter-Systems Interaction units are likely to cost about $200,000. The cost over legacy is therefore the cost of one unit of the system minus the cost of the legacy system. Thus, Warfighter-Systems Interaction will cost $91,000 more per unit than the legacy system.

O&M

AFATDS is an electronics system that is not exposed to the elements, resulting in an O&M percentage of 80 percent. Warfighter-Systems Interaction is designed to have higher capabilities than the legacy system, resulting in higher maintenance cost. The O&M percentage is therefore adjusted upward to 90 percent for the new system. Thus, MOMC is calculated by the standard method at $581 million.

Units to Be Procured

The procurement profile of Warfighter-Systems Interaction is approximated by that of AFATDS. Sixteen years of production history are available for the AFATDS system. Projecting forward to a full 20-year planning horizon for AFATDS, according to the estimation procedure discussed in the section "Universal Production Curves" in Appendix D, suggests a total 6,109 units produced.

ATO#6 Advanced Rotorcraft Technologies

Cost over Legacy

The average unit cost of a Predator UAV is $8.7 million (Nicholas and Rossi, 2007). As a rough estimation in absence of any cost parameters, rotorcraft technology is assumed to add $1 million in unit costs over existing UAVs. Applied to this cost, the extra cost of rotorcraft technology represents an 11 percent price increase.

In total, the production of 900 rotorcraft UAVs represents a $900 million increase over legacy systems.

O&M

Using rotorcraft technology, the operator of UAVs and manned vehicles should be able to switch between winged and rotary modes in mid-flight. This advancement in capability will likely come with a price tag in terms of higher O&M cost. Under the guidelines outlined at the beginning of the appendix, traditional UAVs and manned air vehicles are assigned an O&M percentage of 115 percent. Because of the complex nature of the new technology, rotorcraft vehicles are assigned an O&M percentage of 135 percent. MOMC is calculated by the standard method to be $2.4 billion.

Units to Be Procured

Over a 20-year time horizon, UAVs with installed rotorcraft technologies are likely to replace current medium-range, medium-altitude (tier II) UAVs. These UAVs include Global Hawk, Predator, Reaper, and Pioneer. On the other hand, rotorcraft technology would be unsuited for either tier I or tier III UAV models.[7]

The current and planned inventory for tier II UAVs is as follows. As of 2004, the Air Force planned a total inventory of 51 Global Hawk UAVs ("RQ-4A Global Hawk," 2008). As of 2002, the Air Force also had 50 Predators and planned to obtain 22 more ("RQ-1 Predator MAE UAV," 2008; "MQ-9A Predator B," 2007). An inventory of 18 Reapers is planned by 2009 (Edwards, 2007), and the Navy considers nine Pioneer systems to be a "minimum essential capability" ("Pioneer Short Range [SR] UAV," 2005).

If we assume that each of these numbers is the target inventory for each system, respectively, a total of 150 UAVs with rotorcraft technology would need to be available at any given time.

---

[7] UAVs are categorized into tiers based on the role the aircraft is expected to fulfill. U.S. Army tiers are as follows: tier I, small UAVs; tier II, short-range tactical UAVs; and tier III, medium-range tactical UAVs (Yenne, 2004).

Assuming that the UAV has an operating life of ten years, we estimate that the military will require 300 individual UAVs over the 20-year planning horizon. However, because there are numerous instances of complete loss in battle and routine flight, a factor of two is applied to the above number to arrive at 600 individual units.

Finally, the Rotorcraft Technologies DTO describes the technology as useful for resupply. This is a new feature not currently provided by current UAV technologies, suggesting increased demand for tier II UAVs. As a rough estimate, 300 additional units may be produced over the 20-year planning horizon.

In total, a final estimate of 900 rotorcraft UAVs is assumed.

ATO#7 Theater Effects-Based Operations ACTD

Cost over Legacy

The Theater Effects-Based Operations ACTD is designed to integrate information from large databases in order to provide theater commanders with better intelligence and battlefield awareness. It provides new capabilities and as such does not replace a legacy system. However, it is similar to the Forward Airbase Air Defense (FAAD) Command, Control, and Intelligence (C2I) system, which integrates airbase defense components "into a synergistic system by providing targeting information, air situational intelligence, and information on air/battle management" (Haynes, 1995, p. 52). The cost over legacy for ATO# 7 is based on that of FAAD and is estimated at $14 million (Nicholas and Rossi, 2007).

O&M

Database maintenance associated with this ATO will likely lead to a relatively high level of O&M. An O&M percentage of 80 percent is assigned to the system. O&M is calculated by the standard method to be $324.8 million.

Units to Be Procured

Like the FAAD C2I system, Theater Effects-Based Operations units are likely to be deployed in limited quantity, perhaps with a few units produced for each combatant command in order to monitor large areas of interest. For example, the first Theater Effects-Based Operations system is meant to monitor North Korea.

A procurement history of 14 years was available for the FAAD system (Nicholas and Rossi, 2007). Using this number as a basis, we estimate the projected procurement of Theater-Effects Based Operations systems over a 20-year period to be 29 systems.[8]

---

[8]   See the section "Universal Production Curves" in Appendix D on how the number of new systems to be procured is determined.

ATO#8 Power and Energy

Cost over Legacy

All eight types of Future Combat System (FCS) vehicles scheduled for production are intended to have hybrid engines, which is the focus of the Power and Energy ATO as stated in the DTO data sheet.

As an indication of how expensive hybrid FCS vehicles will be to produce, our estimate is derived from the extra cost of purchasing a hybrid school bus above that of an ordinary one. This extra cost is expressed as a scaling factor of the base cost of the vehicle without a hybrid engine. Assuming that a school bus costs on average $100,000,[9] the complex hybrid technology envisioned under the Power and Energy ATO is likely to initially add $150,000 to this base price, representing a cost increase of 150 percent ("Hybrid Technology," undated). Since the price will fall over the course of the 20-year period, purchases for a hybrid school bus will likely fall to an average of $200,000 per vehicle or lower, roughly twice the price of a regular vehicle.

With this framework in mind, per-unit cost for each of the eight types of FCS vehicles could be estimated in various ways.

The Non–Line of Site Cannon (NLOS-C) is currently under production. The cost of the first units produced averages to $26.2 million (Nicholas and Rossi, 2008). We assume that the cost of the cannon is half of the unit cost. Thus, the vehicle's cost would be the other half of the unit cost or $13.1 million. Similarly, we assume the cost of the Non–Line of Site Mortar to be also $13.1 million.

The Command and Control Vehicle is assumed to be twice as expensive as the M-4 Command and Control Vehicle, which carried a unit cost of $8.675 million (Nicholas and Rossi, 2002), due once again to the cost of hybrid technology. The unit cost is therefore assumed to be $17.35 million ($8.675 × 2).

The Medical Evacuation Vehicle is manufactured with the hull design and chassis of the Bradley M2AO ("Armored Medical Evacuation Vehicle," 2005). The cost of the Bradley M2AO is $1.13 million (Nicholas and Rossi, 2008). We assume that the hybrid technology integrated into the Medical Evacuation Vehicle about doubles this per-unit cost to $2 million.

The costs of the Recovery and Maintenance Vehicle and the Reconnaissance and Surveillance Vehicle were unavailable to us. However, we know that these vehicles have similar crew configurations, speeds, and ranges as the medical evacuation vehicle. Therefore, we assume that the cost of these vehicles will be $2 million ("Manned Ground Vehicle," 2005).

The Infantry Combat/Carrier Vehicle is a troop carrying vehicle similar to the Stryker. Because the Stryker is already more complex than other military vehicles and thus carries a higher per-unit cost, we assume that the introduction of hybrid technol-

---

[9] The price of a traditional bus was reported to be $95,000 (Hoffman, 2007).

ogy will not double the cost of this vehicle. Rather, whereas the Stryker costs $3.281 million per unit (Nicholas and Rossi, 2008), we assume that the Infantry Combat/Carrier Vehicle will cost 1.5 times this amount, or $4.92 million.

The Mounted Combat System (MCS) is envisioned as a replacement for the M1 Abrams and M2/M3 Bradley vehicles. On average, these two legacy systems cost $1.7 million per unit (Nicholas and Rossi, 2008). Assuming that hybrid systems increase the cost for such a vehicle by 100 percent, we assign the MCS a cost of $3.4 million per unit, or $1.7 million times two.

The estimated average cost of FCS vehicles is $7.27 million. However, this number is skewed upward by more expensive cannon units, which are likely to see relatively less production. Using a more appropriate mix of vehicles, average cost is estimated to be $5.62 million per vehicle.[10] Since the FCS vehicles are assumed to be twice as expensive as legacy vehicles, the cost of a legacy fleet of vehicles is assigned a cost of $2.81 million ($5.62 × 1/2). As a result, cost over legacy for the system is also $2.81 million ($5.62 – $2.81).

O&M

The legacy vehicles replaced under this ATO were assigned an O&M percentage of 115 percent, based on the high level of operating expenses and maintenance needed for vehicular systems. The FCS vehicles are constructed using a common drive train resulting in maintenance savings. Moreover, lower fuel consumption resulting from the hybrid engine also lowers O&M cost. In total, the O&M of the new system is assumed to be 95 percent. The MOMC calculation for this ATO has been discussed above, and calculations are based on the acquisition cost of the legacy system. As a result, MOMC falls significantly and ends up at negative $9.4 billion (i.e., savings in O&M cost when new systems are used instead of legacy systems).

Units to Be Procured

Reports suggest that over the next 20 years, 15 brigades will be fully outfitted with FCS vehicles. Each brigade will receive on the order of 700 vehicles, resulting in a total of 10,500 vehicles fielded ("Boeing Project Takes Aim," undated). Other brigades may receive a limited number of FCS vehicles to increase capabilities. No estimates are available to suggest the number of vehicles provided to each of the other 27 Army brigades. Assuming they will receive one-third as many vehicles as fully outfitted brigades, 6,300 further vehicles will be fielded, yielding a total of 16,800 vehicles fielded in the first 20 years of the FCS rollout.

A rough approximation suggests that the bulk of the vehicle mix for a typical brigade would be made up of troop carriers and combat vehicles, with relatively fewer

---

[10] Our estimation of the number of units and mix of FCS vehicle types appears under the "Units to Be Procured" subsection for this weapon system.

cannons, command and control vehicles, and other specialty vehicles, such as the recovery and maintenance vehicle. In order to develop units to be procured and MUC estimations, we assume the following scheme as to the vehicle types that would make up the 700 FCS units per brigade: 30 each of the NLOS-C and NLOS-M (non–line of site cannons and mortars, respectively), 50 each of the command and control vehicle and recovery/maintenance vehicle, 45 each of the medical and evacuation vehicle and reconnaissance and surveillance vehicle, 300 of the infantry combat/carrier vehicle, and finally 150 of the MCSs.

ATO#9: Future Tactical Truck System

Cost over Legacy

The DTO of the Future Tactical Truck System is to improve truck performance by reducing maintenance requirements and fuel consumption, and will be applied to all truck classes, including HMMWVs as well as the family of Medium Tactical Vehicles and the family of Heavy Tactical Vehicles. Based on the DTO description, production costs are estimated to rise 15 percent compared to current costs.

The current average cost of all truck classes is $151,000 per unit (Nicholas and Rossi, 2007). That number is expected to rise to $173,650 based on a 15 percent cost increase, representing a cost over legacy of $22,650 per unit.

O&M

Because lower maintenance is specifically mentioned in the DTO description of the system, the O&M percentage is expected to fall from the default 115 to 105 percent. MOMC is calculated as in the case of the Power and Energy ATO—the DTO description specifically indicates that this weapon system will reduce O&M costs. We estimate that these costs will fall by $2.8 billion compared to the current legacy system.

Units to Be Procured

This ATO replaces three vehicle classes. The production history for each is available for the last 20 years. Added together, total production of these historical systems approximates the future production of improved trucks, yielding 182,483 total vehicles (Nicholas and Rossi, 2007).

ATO#10 Joint Precision Airdrop System

Cost over Legacy

Current airdrop capabilities feature a parachute and not much else, suggesting a legacy system with negligible production cost. On the other hand, the Joint Precision Airdrop System introduces precise navigation components into airdrop delivery.

The Longbow Hellfire missile is the system we used for cost comparison, because it is equipped with an active-seeker, air-to-surface capability. The cost of one missile is $161,000 (Nicholas and Rossi, 2007). Navigation components of the airdrop system do not have to be as intricate as those on a fast-moving missile which tracks a moving target. The cost over legacy of one airdrop system is assumed to be $150,000.

O&M

This system contains electronic components and is subject to outdoor use. It is assigned an O&M percentage of 115 percent. O&M is calculated by the standard method to be $172.5 million.

Units to Be Procured

The capabilities developed under this DTO provide for precise airdrop from very high altitude so as to avoid hostile fire on delivery aircraft as well as to ensure appropriate placement of re-supply. Currently, the Air Force C-130 fleet consists of about 500 such aircraft.[11] As an approximation, we assume one airdrop system for each plane.

Electronic navigation components are likely to be housed in such a manner as to prevent major breakdown. However, units lost following airdrop because of ground impact and other accidents may be high, suggesting an operating life of ten years and total production of 1,000 units over 20 years.

ATO#11 Warfighter Physiological Status Monitoring

Cost over Legacy

The DTO develops a body-worn monitoring sensor with minimal size and weight to measure organic compounds, analytes, and psychological parameters so as to optimize performance during stressful physical activities and combat (U.S. Department of Defense, undated). The monitoring system produced under this ATO provides new capabilities that do not replace those of any existing system, so that the cost of procurement for the new systems is the same as the cost over legacy of the systems.

---

[11] The Air Force inventory for C-130s: active force, 186; Air National Guard, 222; and Air Force Reserve, 106 (U.S. Air Force, 2008).

To be viable for widespread use, the Warfighter Physiological Status Monitoring system will have to be relatively inexpensive, given the number of units to be produced. We assume the unit cost to be $500.

O&M

Electronic systems experiencing heavy outdoor use are assigned an O&M percentage of 115 percent. This system is assigned an O&M percentage of 105 percent to account for its simplicity relative to other, more complex electronics systems. MOMC is calculated by the standard method as $98 million.

Units to Be Procured

The sensors produced under this ATO will be distributed to troops only during contingencies. As of mid-December 2006, 152,000 troops ("U.S. Forces Order of Battle," 2008) were deployed in Iraq, while as of 2004, 17,900 additional troops were deployed in Afghanistan, totaling 169,900 troops in hostile environments (Burges, 2004). Other contingencies might add 10 percent to this number, totaling 182,483 systems needed.

However, during contingencies not all troops are exposed to potential combat at any time. Perhaps half of the troops deployed in a contingency would require a monitoring system at any given time, suggesting a necessary production of 93,445. We further assume one replacement over the 20-year period will be needed. Thus, the total number of units procured will be 186,890.

ATO#12 Battlefield Treatment of Fractures and Soft Tissue Trauma Care

Cost over Legacy

The main technology produced under this ATO is a dressing that delivers pure oxygen to wounds in order to reduce infection. No source was available to indicate the cost of the new dressing. Assuming that new wound dressings will cost $5 versus $1 per unit for ordinary dressings, the cost over legacy for the system is $4.

O&M

O&M cost is not expected to change between the old and new wound dressings, although the new dressing has added complexity. There is no indication that it requires refrigeration or other special storage conditions. MOMC is set to zero.

Units to Be Procured

The number of wounded soldiers in Iraq over the course of the current conflict is used to estimate the number of wound dressings needed over the 20-year planning horizon. So far, 30,324 wounded soldiers have been recorded in Iraq (icasualtics.org, undated).

If we assume five patches must be kept in inventory for each wounded soldier, 151,620 dressings would be needed per conflict. However, we assume that future conflicts will last two rather than five years, resulting in a reduced estimate of 60,648 dressings. If these patches are consumed or expired in ten years, 121,296 patches will be needed. Finally, 10 percent more units will be used in non-contingency situations, bringing the total number of wound dressings used to 133,426.

ATO#13 Intravenous Drug to Treat Severe and Complicated Malaria Caused by Multidrug-Resistant Malaria

Cost over Legacy

Over a 20-year time horizon, when the new treatment will become commonplace, the cost of the new treatment developed under this ATO can be similar to the current treatment cost for malaria. Cost over legacy is therefore assumed to be zero.

O&M

Costs are not expected to rise above those associated with the legacy system. MOMC is set to zero.

Units to Be Procured

Unit estimation does not need to be performed in this case because cost over legacy is zero and O&M does not change between new and old systems.

ATO#14 DNA Vaccines to Prevent Hemorrhagic Fevers

Cost over Legacy

One dose of vaccine for the prevention of dengue (a type of hemorrhagic fever) costs $20 in the private sector, but only $1 in the public sector (Shepard, 2004, pp. 9–10). DNA vaccines may be harder to preserve and deliver than traditional vaccines, but over a 20-year time horizon, costs between the two should not differ by a very large magnitude. We assume an average cost of $5 per dosage for DNA vaccines, compared to $1 for regular vaccines. Because no vaccination is currently available to inoculate against hemorrhagic fevers, cost over legacy system is also simply $5 per dosage.

O&M

DNA vaccines do not require refrigeration or other special storage. As a result, they are assigned an O&M percentage of 10 percent. MOMC is calculated by the standard method to be $858,000.

Units to Be Procured

The estimated quantity of DNA vaccine doses utilized by the Army will depend on the number of doses per individual treated over a 20-year time horizon. Recent trials with a DNA vaccine conferred immunity for nine months against the Ebola virus (Touchette, 2004). Assuming that a comparable vaccine could be developed for hemorrhagic fever based on DNA, one dose would probably be enough to provide immunity from disease over an entire tour of duty, which on average lasts 320 days (Bennis, 2004). As a result, roughly one dose a year should be adequate for each treated soldier.

Only service members who serve in regions at high risk for disease are likely to need this vaccine. Available numbers suggest that 170,000 troops are stationed in Iraq and Afghanistan and surrounding areas of Southwest Asia ("Surviving a Tour of Duty," 2005). It is unclear how many of the four contingencies assumed for the 20-year period will occur in "at-risk" areas or how many years such a contingency will last. We assume four contingency years in "at-risk" areas to yield a requirement of 680,000 doses. In addition, there are about 45,000 troops in other at-risk areas of Asia (excluding Japan),[12] 6,000 in other areas of the Middle East, and 800 in Africa. Thus, there is a requirement of 51,800 doses per year or 1,036,000 doses over the 20-year period. The sum of these two requirements is 1,716,000.

ATO#15 Fatigue Management Capability for Sustained Readiness and Performance

Cost over Legacy

This DTO will only produce software, which can be distributed at negligible production cost. No cost over legacy estimate is needed.

O&M

No O&M difference is expected between any old and new systems. Software in general is not considered to require an O&M expense. MOMC is set to zero.

Units to Be Procured

Based on the above discussion, no production estimate is necessary for this system, as the MOMC is zero.

---

[12] The numbers of troops here and below in this paragraph come from *GlobalSecurity.org* as of November 24, 2007.

ATO#16 Physical Training Interventions to Enhance Military Task Performance and Reduce Musculoskeletal Injuries

Cost over Legacy

No specific equipment is produced as a result of this DTO. Instead, the DTO develops training and monitoring methods, which do not suggest different production or manpower costs in relation to current methods.

O&M

No change in O&M cost is anticipated. Costs are also identical to the legacy system, and MOMC is zero.

Units to Be Procured

No estimate is necessary, as cost over legacy or MUC is zero.

ATO#17 High-Altitude Airship

Cost over Legacy

High-Altitude Airships provide a surveillance capability currently not satisfied either by satellites, which do not provide the same resolution quality, or by surveillance aircraft, which cannot provide the same continuous surveillance capability over a long period of time. Lockheed Martin, the producer of the Airship, "hopes to keep the unit cost of the operational airship at roughly $50–60 million before its advanced radars, sensors et al." ("Lockheed Wins $149.2M Contract," 2006). We use an average of $55 million as the probable cost of production per unit.

O&M

This system does not require constant manning in normal operation and is subject to less physical wear compared to other air systems. As a result, it should be slightly less costly to maintain, and is assigned an O&M percentage of 105 percent. MOMC is calculated by the standard method at $2.89 billion.

Units to Be Procured

The high-altitude, long-deployed nature of these surveillance airships suggests the production of few units used to protect critical areas such as airbase and the Green Zone during contingencies. Currently within Iraq, roughly 25 such areas might be necessary for surveillance. Another 25 might be deployed in other areas, including Afghanistan and South Korea. Total production is thus 50 units. Because the system is at a very high altitude and not likely subject to hostile attack as a result, the operating life of all units is 20 years, with no expected loss during operation.

ATO#18: Vaccine for Prevention of Military HIV Infection

Cost over Legacy

Over a 20-year planning horizon, DNA vaccines are assumed to cost $5 per dose, same as the DNA vaccines for the prevention of malaria in ATO#2. This system does not replace the capability of any legacy system, and therefore the MUC is also $5.

O&M

Like the other vaccines discussed in this appendix, the O&M percentage is 10 percent. MOMC is calculated by the standard method to be $1 million.

Units to be Procured

The HIV vaccine developed is intended to prevent military personnel from being permanently non-deployable. Rather than a vaccination given only to deployed troops, it will be a general vaccination for all Army members. Currently, there are 507,082 active duty and 333,177 guard troops in the Army. A further 189,005 in reserve suggests total troop strength of 1,029,264.[13]

Assuming that one dose of the vaccine is effective for the service life of each military member and that there is a turnover of 5 percent a year in service members, we estimate that, over a 20-year time horizon, 2,058,528 troops will be inoculated with the vaccine.

ATO#19: Joint Rapid Airfield Construction

Cost over Legacy

The Joint Rapid Airfield Construction techniques developed in this DTO introduce stabilizing techniques and appropriate addition of substances to soil so that temporary airfields last longer. While the equipment used to analyze airfields and substances used to stabilize soil may represent an additional expense above current temporary airfield construction techniques, the savings in maintenance costs may end up with a zero net change in this cost component.

On the other hand, personnel costs associated with rapid airfield construction are likely to be significantly lower than prior methods. Ten to 12 soldiers can do what previously required 30 (U.S. Army Corps of Engineers, undated). Using the new rapid construction techniques, construction time for adding two parking aprons to a temporary airfield was reduced from three weeks to three days (Anderton, 2005).

The old construction method requires 21 days of labor by 30 troops (Anderton, 2005). At a labor cost of $274 per person per day (which in turn is the daily average

---

[13] All numbers in this paragraph come from the following source: Smith, 2007.

cost if wages are $100,000 per year), construction costs to build one airfield using the legacy system comes to $173,000. Using the new system, 12 troops work for three days, resulting in a per-airfield cost of $9,900. The savings between the old and new systems is $163,000 per airfield. In the course of constructing eight airfields per contingency and four contingencies over 20 years, this represents a total savings of $5.21 million.

Military trucks in general cost between $150,000 and $200,000,[14] but the RAVEN necessary for airfield construction will be specialized heavy equipment. Adding a premium as a result, we assume that each RAVEN necessary for rapid airfield construction costs $350,000. As stated above, 32 RAVEN vehicles would have to be produced, at a total cost of $11.2 million over 20 years.

Subtracting the savings in labor expenses from the cost of the RAVEN vehicles produced yields a total cost over legacy of $5.99 million for this ATO.

O&M

We assigned an O&M percentage of 80 percent for operating and maintaining the legacy system for rapid airfield construction. In this case, the RAVEN vehicles produced under this ATO will increase that figure. The O&M percentage is assumed to increase to 100 percent, representing a net change of 20 percent. MOMC is calculated by the standard method to be $7.1 million.

Units to Be Procured

We assume that there will be four major contingencies over the next 20 years for which systems derived from the 29 ATOs will be deployed. Each contingency will require rapid airfield construction techniques a total of 32 times for eight temporary airfields.

The joint rapid airfield construction techniques require the use of a specialized construction vehicle called RAVEN (Rapid Assessment Vehicle Engineer). Because rapid airfield construction is so rare, an inventory of 16 RAVENs would likely satisfy need at any given time. Because RAVENs are involved in a heavy-use activity, we further assume that the operating life of a RAVEN is only ten years. Over a 20-year planning horizon, 32 RAVENs would need to be procured. In the above section on cost over legacy, we have estimated all 32 RAVENS and labor savings in rapid airfield construction as one unit that cost $5.99 million.

---

[14] For example, Army tactical vehicles have an average cost of $173,650 per vehicle (Nicholas and Rossi, 2007).

ATO#20: Dynamic Mission Readiness Training for C4ISR

Cost over Legacy

The training developed under this DTO is supposed to produce a savings in training time for operational planners of one week in each year over the legacy system, and one and a half weeks for intelligence officers over the legacy system.

Assuming a ratio of four operational planners to every intelligence officer, we define NU as the total number of people trained (operational planners + intelligence officers) and WS as the weekly salary of trainees. Then the cost savings per unit can be expressed as:

$$\text{Cost savings per person trained} = [(4(NU) \div 5) \times (WS) + ((NU) \div 5) \times (3(WS) \div 2) \times 1.2] \div NU,$$

where WS = ($100,000 ÷ ·52), and the salary of an intelligence officer is assumed to be 20 percent higher than that of an operational planner.

The above equation simplifies to:

$$\text{Cost savings per person trained} = (\$100,000 \div 52) \times (11.6 \div 10) = \$2,200 \text{ in total savings per person trained.}$$

Over 20 years, 1,000 persons will be training and a savings of $2.2 million will result.

O&M

This ATO is assigned the lowest O&M percentage of 10 percent. However, O&M cost does not increase over the legacy system, therefore MOMC is equal to zero in calculations.

Units to Be Procured

This DTO develops training for operational planners and intelligence agents. The number of such service members trained each year is likely to be low. As a very rough estimate, assume that 50 such individuals are trained each year for a total of 1,000 individuals over 20 years.

ATO#21 Future Force Warrior ATD

Cost over Legacy

The system suggests the types of and number of devices to place on particular soldiers within platoons in order to optimize performance. However, producible devices are not developed within this DTO, and as such the project does not result in any changes to existing production.

O&M

No O&M estimate is associated with this ATO.

Units to Be Procured

No estimate is necessary.

ATO#22 Leader Adaptability

Cost over Legacy

This DTO is to improve training performance, but it does not seem to represent any change in man-hours used in training and does not seem to produce any new, physical training materials that might incur production costs. Any marginal cost associated with the ATO is therefore negligible.

O&M

No estimate is necessary.

Units to Be Procured

No estimate is necessary.

ATO#23 Advanced Antennas

Cost over Legacy

This ATO attempts to drastically transform and improve Army communications capabilities in step with the introduction of FCS. New antennas increase maximum data throughput across a wider range of environments. They are designed to allow conformal integration in vehicle platforms and clothing. They must possess these capabilities while providing on-board and on-person power sources at reduced size and visual signature. And they must achieve these goals at a minimum cost.

The broad goals outlined under this ATO mean that multiple weapon systems and weapon components are replaced, including body-wearable antennas and satellite receivers. The complexity of the ATO is beyond the scope of this method demonstration. As a result, discussions and calculations related to advanced antennas focus exclusively on the provision of body-wearable antennas for the FCS, since FCS provision estimates had already been carried out for another ATO ("Power & Energy"). In that regard, while performance improvement is expected, costs are specifically mentioned as being roughly equal to those of the legacy system as a result of production efficiencies associated with manufacture of the improved technology.

O&M

Nothing described in the description of this system suggests a change in O&M costs. Thus, MOMC is zero.

Units to Be Procured

No estimate is necessary.

ATO#24 Autonomous Distributed Sensors

Cost over Legacy

This system does not replace a capability already provided by any existing system, and the new system provides long-term, passive detection of threats.

The SQS-53b is a submarine detection sonar for surface warships with similar characteristics to the Autonomous Distributed Sensors systems. Each SQS-53b system costs $5 million (Nicholas and Rossi, 2007). The SQS-53b system features both active and passive detection, whereas the autonomous sensors in this ATO provide only passive detection. Moreover, the SQS-53b sensor system is integrated with a helicopter to respond to attack, which is not the case with autonomous sensors in this ATO. Finally, SQS-53b sensors are for use under water, where background noise may make detection of threats harder and require a much more sophisticated sensor than necessary under the autonomous sensor system.

Based on these differences, the cost of autonomous distributed sensors is assumed to be about 40 percent below that of the SQS-53b system, resulting in a cost of $3 million per sensor system produced.

O&M

These are electronics systems subject to outdoor use, and are therefore assigned an O&M percentage of 115 percent. MOMC is calculated by the standard method to be $1.8 billion.

Units to Be Procured

Autonomous, self-powered sensors are useful for detecting attack and protecting large areas of potential vulnerability where electricity is unavailable. During the initial stages of a contingency, autonomous sensors can be used to detect ambushes on U.S. forces at the front. As of November 2006, 152,000 troops were deployed in Iraq ("U.S. Forces Order of Battle," 2008). If 10 percent of that number is actually deployed on the front lines, 15,200 troops will need to be protected on a front. If we further assume two fronts during a contingency, 30,400 troops will be vulnerable to frontline attack. If one sensor system is needed per 100 soldiers, 304 systems will be deployed during contingencies. Finally, if the operating life of the sensor is ten years, 608 total sensor systems will be produced.

ATO#25 Tactical Wireless Network Assurance

Cost over Legacy

Entrust, a network security provider, spends approximately $118 million to provide security to 1,500 different companies (*The Value Line Investment Survey,* 2007, p. 1446).[15] Assuming an average of 500 people at each of these companies, security can be provided to 750,000 separate individuals for this sum. On a per-person, per-year basis, network security therefore costs about $157 to provide.

Based on the need estimated in the subsection "Units to be Procured" below for the 20-year planning horizon, costs during contingencies will come to $157 per person for 200,000 total troops, or $31.4 million in total. During peacetime operations, costs each year will come to $1.18 million, based on 7,500 troops utilizing tactical networks each year. Over a 20-year period, this comes to a total cost of $23.6 million. In total, between peacetime and contingency use, tactical wireless network assurance is assumed to cost $55 million, which is also the estimated cost over legacy given no legacy system in place.

O&M

The operation and maintenance of networks is considered as part of cost over legacy in the above calculations. Therefore, no separate O&M cost needs to be estimated.

Units to Be Procured

For the military, tactical wireless networks will be used during major contingency operations as well as during peacetime operations over the next 20 years.

We assume that there will be four major contingencies over 20 years. During the initial phases of any contingency (assuming to last for the first year), 50,000 frontline troops will need to be supplied with a secure network ability, assuming that 100,000 Army troops are involved in initial operations near the front, but only half are mobile troops in the field and would require tactical network capabilities. Thus, for four major contingencies, a total of 200,000 soldiers would need the wireless network device for one year. Moreover, during peacetime operations, 7,500 troops are assumed to be involved in patrols in any given year that might require tactical wireless networks.

---

[15] We used the projected revenues of four quarters ended on December 31, 2007, and March 31, June 30, and September 30, 2008.

ATO#26 Command, Control, and Communications On-the-Move Demonstration

Cost over Legacy

This demonstration does not develop unique hardware. Rather, it seeks to develop techniques to effectively manage and maintain on-the-move communications. As a result, the DTO is likely to result only in RDT&E costs, but not change or introduce new production or manpower costs of any kind. Because RDT&E costs are treated separately and not considered as part of the cost of production in this appendix, cost over legacy is zero.

O&M

No estimate is necessary.

Units to Be Procured

No estimate is necessary.

ATO#27 Modeling Architecture for Technology, Research, and Experimentation

Cost over Legacy

The DTO increases the ease with which research models can be constructed and analyzed. In general, increased efficiency reduces costs. On the other hand, rapid changes in models being analyzed can result in the need for more manpower and an increase in work-related communications. We assume that there is no net change in procurement cost from legacy systems to new systems.

O&M

No O&M costs are associated with this ATO.

Units to Be Procured

No estimate is necessary.

ATO#28 Radio-Enabling Technologies and Nextgen Applications— Joint ATD

Cost over Legacy

The radio developed under this ATO has a signal range of 30 megahertz to 2.8 gigahertz. The span from the low end of this range to the high end is similar to the frequency profile of the model KAW2180 power amplifier, which runs from 100 kilohertz to one gigahertz.

The KAW2180 costs $32,000 (Nicholas and Rossi, 2007). Since the amplifier produced under the DTO is intended as an improvement over currently available products, we assume that the amplifier will be produced at a higher cost, about $40,000 per unit. As a result, cost over legacy is estimated as $8,000.

O&M

This is a fielded electronic system with more complex components than the legacy system it replaces. Adjusting slightly upward from the default O&M percentage of 115 percent, we assign this system a percentage of 125 percent. MOMC is calculated by the standard method to yield $88 million.

Units to Be Procured

This ATO develops a man-pack radio. The Army will have to maintain enough units in order to fully supply all troops during the initial stages of a contingency. Assuming that 50,000 troops are necessary during this initial stage and further assuming four contingencies over the next 20 years, we estimate as a baseline that the military will have to maintain 50,000 units at any given time. However, since the radio is a large man-pack, it will likely be carried by one specialized soldier in a group of 15, rather than as a personal radio for each individual. Dividing 50,000 by 15, we project a necessary supply of 3,333 units at any given time.

Assuming that the operating life of the radio is ten years, we estimate that 6,666 units will have to be purchased over the 20-year planning horizon.

ATO#29 On-Route Detection of Mines/IEDs

Cost over Legacy

This DTO is for integration of components, not hardware development. We assume that there is no increase in production costs.

O&M

There is no indication that the O&M cost for the new system will be higher than that for the legacy system. We assume MOMC to be zero.

Units to Be Procured

Because cost over legacy is zero and no additional O&M costs are associated with this system, the number of units produced does not need to be estimated.

Universal Curves for Estimation of the Remaining Lifecycle Cost for Components

Over the lifespan of a weapon system, RDT&E costs are broken into three components. Prior to production, S&T and SDD occur in sequential order. Following production, there will be an upgrade cost, which includes the cost for correcting minor design defects and the cost for improving performance. The purpose is to allow the basic design of the weapon system to last longer and to postpone the need for starting a new ATO for the next generation of the system or a completely new system.

In this appendix, we detail the approach to estimate the total S&T cost, the SDD cost, and the upgrade cost. We also describe the procedure to estimate the number of units of an ATO-derived system to be fielded over 20 years.

Total S&T Cost

For S&T cost, the model requires only remaining S&T cost that is to occur in the future.[1] Past S&T cost, which is a sunk cost, plays no role in portfolio management including decisions concerning the future of individual projects. Thus, we introduce total S&T cost here not to use it in the model directly, but rather to project the SDD cost and the upgrade cost, because the projection is based on the correlation among the three cost components (S&T, SDD, and upgrade) in existing systems that are similar to the ATO-derived system in question.

Historical S&T

In order to calculate total S&T, we first had to record historical S&T, that which occurred prior to FY 2005. However, the historical S&T data for a specific ATO may be missing or incomplete. Our budgetary information comes from the DTO data

---

[1] As explained in the subsection "Estimation of Cost Components" in Chapter Four, we simply looked up the future or remaining annual S&T costs from the DDR&E Web site.

sheets provided by the DDR&E. In this data source, the yearly budget is first broken down into program elements (PEs),[2] which are further segregated into "projects," which in turn are broken down into specific "areas of research" under each project. Thus, a single PE provides funding to multiple areas of research. Consider ATO 2, Vaccines for the Prevention of Malaria. Army funding for this ATO falls under two PEs, one of which is 0602787A. Under this PE number, the specific funding "project" is 870. Turning to the DDR&E data sheets, we find that project 870 currently provides funds to the following areas of research:

- malaria vaccines
- antidiarrheal vaccines
- insect control
- scrub typhus vaccine
- vaccines against dengue fever
- malaria drug candidates.   .

Clearly, it is necessary to have a line-item description of each of these areas of research in order to parse out the part of the budget that belongs exclusively to ATO 2, Vaccines for Prevention of Malaria. However, research area descriptions are often vague, opening the door to coder error. In the list above, "malaria vaccines" would be the most likely candidate as a match for ATO 2. However, one wonders if some S&T funding for ATO 2 is also being recorded under "malaria drug candidates." A more serious problem is that as one goes back in time, budget descriptions are less and less detailed. Turning again to our example, the FY 2004 data sheets are the first to include descriptions of each funding area under project 870. Furthermore, the FY 2004 budget sheets contain data going back to 2002. On the other hand, FY 2003 budget sheets contain the aggregate funding for project 870, but they do *not* break this down into funding areas. Thus, actual funding for 2001 and earlier is not available. If an ATO takes up all of the funding of a project, we can still use these data. In the case of ATO 2, however, no individual budget data are available prior to 2002. Finally, note that data sheets are only available back to FY 1999. These data sheets provide information back to 1998. Budget data prior to this year are available only at the program element level and are not available for specific projects or areas of research.

Given the above discussion, we found the first year of historical S&T data available for each ATO to be the following:

- *FY 1996.* The ATO has its own exclusive PE number, with no other projects or areas of research falling under this PE.

---

[2] The PE is the building block of the Future Years Defense Program. The PE describes the program mission and may consist of forces, manpower, materiel, services, and associated costs, as applicable.

- *FY 1998.* The ATO has its own exclusive project under a PE number, with no other areas of interest falling under this project.
- *FY 2002.* The ATO is an area of research under a PE.

Remaining S&T

Since the 29 ATOs were selected from the FY 2005 list,[3] we tried to simulate the decisionmaking of the portfolio manager from the vantage point of the end of FY 2005. The remaining S&T budget available at the time would be that appearing in the DTO data sheets and the *RDT&E Descriptive Summaries* for FY 2006.[4] However, the FY 2006 data would show annual budgets only through FY 2011. Thus, one might not know at the time (end of FY 2005) whether any of the 29 ATOs would be funded beyond FY 2011. Moreover, the budgetary projection may end earlier, say, in FY 2009. This does not necessarily mean that the ATO ends in FY 2009—it may mean only that no projection was made beyond FY 2009 at the time (the end of FY 2005). However, because there is no systematic way to determine whether an individual ATO will receive additional S&T funding beyond that already apportioned, for this study we simply assume that the funding information recorded in the DTO data sheets was complete and accurate.

Calculating Total S&T

As stated above, the funding for a given ATO is described under one or more PEs. Our approach to calculating historical S&T is to identify and/or estimate the full funding stream for a given ATO under each of the pertinent PEs. Then, we add up all of the streams to yield the full funding stream for the ATO. This number is added to the remaining S&T budget found in the DTO data sheets in order to calculate total S&T.

System Development and Demonstration Cost Curve

In this study, we focused on ATOs, which are established during 6.2 and 6.3 programs. Our approach is to project SDD cost for each ATO based on its cost for 6.2

---

[3] There is a time lag in the availability of a published list of the latest ATOs. The FY 2005 list was the latest list available at the time of this study. Should the Army S&T community wish to apply our method and model, they can use their up-to-date list of ATOs.

[4] The DTO data sheets are shown in a DDR&E report (DDR&E, 2006). However, the *RDT&E Descriptive Summaries* are published annually by the Defense Contract Management Agency.

and 6.3 programs. The first step is to trace back all the historical funding streams for the 6.2 and 6.3 programs for each ATO. This step is accomplished by identifying all the pertinent Army PEs. Adding the historical cost to the remaining S&T cost yields the 6.2/6.3 program cost for each of the 29 ATOs.

The second step is to determine a cost relationship among 6.2/6.3 programs and the SDD cost. To project the SDD costs for the 29 ATOs, we turned to existing weapons that are similar to the ATO-derived systems and assumed that the correlations between the 6.2/6.3 and SDD of these similar existing systems are the same as those for the ATO-derived system in question. In other words, knowing the 6.2/6.3 cost of an ATO-derived system and using the same 6.2/6.3 correlations with the SDD cost for the existing systems, we estimated the SDD cost for the ATO-derived system in question.

Specifically, we used the historical RDT&E[5] data for existing weapon systems that were still being produced and procured as of 2006. From the data, we estimated the proportion of historical expenditures that applied to 6.1, 6.2/6.3, and SDD costs, respectively. We chose existing electronics systems as the similar systems to the ATO-derived systems.[6] We found 17 years of aggregate historical data for 25 electronics systems.

RDT&E funding prior to any production of systems can be considered S&T (6.1 to 6.3 programs) and SDD funding, chronologically. Our data source did not provide any indication as to what proportion of spending was 6.1, 6.2/6.3, and SDD, respectively, for each weapon system. We assume that the 6.1 basic research program takes two years to complete and exclude it for the development of the cost relation between 6.2/6.3 and SDD. We then split each funding stream prior to any production into two equal time periods. The funding for the first half was considered 6.2/6.3 programs, and that for the second half was considered SDD. We normalized the first year of 6.2/6.3 data to equal one, so that the absolute magnitude of funding for any particular weapon system does not lend it more weight than other systems in determining the appropriate ratio between 6.2/6.3 and SDD. Then, we expressed the ratio of the second half of funding (assumed to be SDD) to the first half of funding (assumed to be 6.2/6.3) for each weapon system as a multiplier, and we averaged across all 25 weapon systems with any historical RDT&E funding prior to production, giving us a ratio of 1.44 (Figure D.1). In other words, for the 25 historical electronics weapon systems still in production as of FY 2006 and for which data were available, SDD funding is 1.44 times the level of total 6.2/6.3 funding. We simply multiply our total 6.2/6.3 funding projections by this ratio to find projected SDD levels.

---

[5]   We assumed that the RDT&E cost incurred after the fielding of the weapon system is the system upgrade cost.

[6]   These electronics systems appear in Nicholas and Rossi, 2006. We also used earlier editions to trace the historical costs to the beginning of their RDT&E.

Figure D.1
SDD to S&T Ratio for Current Electronics Systems

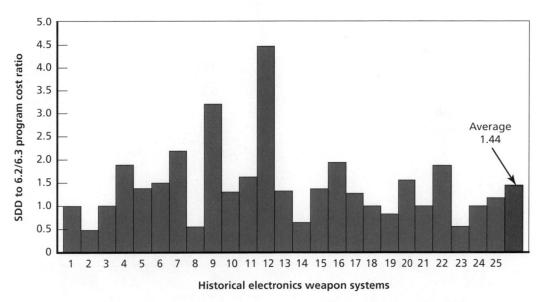

RAND *MG761-D.1*

Upgrade Cost Projections

Because data related to upgrade costs for weapon systems were not available to us, several assumptions had to be made for our estimate. First, upgrade costs are assumed to be recorded as part of RDT&E for historical weapon systems in our data source, *U.S. Weapon Systems Cost* (Nicholas and Rossi, 2002, 2006, 2007, and 2008). Second, upgrade costs are assumed to occur after first procurement of a weapon system and run concurrently with subsequent procurements. Thus, all RDT&E data recorded during and after the first year of recorded procurements in our data source are assumed to be upgrade cost. We have assembled a complete history of upgrade cost data for 21 historical electronics systems.[7] The relationship between upgrade cost and total procurement cost is shown in Figure D.2. Several outliers notwithstanding, systems with high total costs tend to have high upgrade costs. A linear relationship between upgrade and total procurement cost is also shown in the figure.

---

[7] We did not use systems that either were still in production, so that not all future upgrade cost is recorded for the system, or were older than the beginning of our recorded data, so that not all historical upgrade cost is recorded.

Figure D.2
Relationship Between Upgrade Cost and Total Procurement Cost for Historical Electronics Weapon Systems (in millions of dollars)

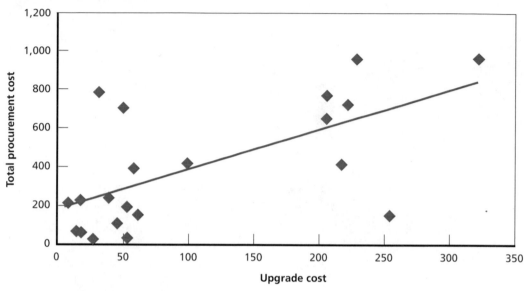

NOTES: y = 2.0416x + 176.53. R^2 = 0.4071.
RAND *MG761-D.2*

From the slope of the fitted straight line, one can see that the upgrade cost is about half of the total procurement cost. To get a better approximation, we re-plotted the data as a bar chart in Figure D.3 and determined the average ratio (in red) to be 0.406, as opposed to about 0.5 from the slope in Figure D.2. For the 29 new ATOs considered in this study, upgrade cost is equal to the acquisition cost of the ATO times the average historical ratio of 0.406. As an example, ATO#3, Overwatch, carries an estimated acquisition cost of $6.25 billion. Upgrade cost is therefore projected to be $2.5 billion ($6.25 × 0.406).

Universal Production Curves

The estimated units produced for each weapon system were obtained from one of two sources and applied over a 20-year time horizon.

If no similar system exists for an ATO-derived weapon system, we estimate units produced based on likely demand over 20 years, taking into account the likely geographical distribution of systems needed in the future, the likelihood of use in contingencies, useful life of units produced under the system, and other pertinent information.

Figure D.3
Upgrade Cost to Total Procurement Cost Ratio for Historical Electronics Weapon Systems

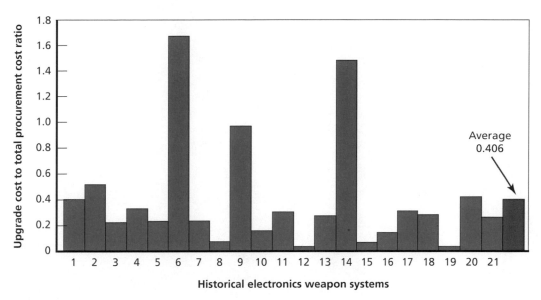

In other cases, we match the ATO-derived system to an existing system or systems. For example, a radio to be developed under the ATO on Small Unit Operations TD may replace SINCGARS, or have a similar production profile. If so, the total SINCGARS units produced over the last 20 years can be assumed to be the total likely units produced for this ATO-derived radio. The production data from SINCGARS were taken from *U.S. Weapon Systems Costs* (Nicholas and Rossi, 2006) and treated as the likely production schedule of this ATO-derived radio.

When a 20-year history of production was not available for the similar existing system, we used a production curve derived from the available production history of all historical weapon systems of the same class. We have developed a universal curve for any ATO-derived electronics system from aggregate historical data for 34 electronics systems. This universal curve was produced in the same way as were other projections described in the above two sections. The first three years of data are averaged, and then the numbers of units produced in subsequent years are divided by this number. Each year of production is therefore expressed as a multiplier of the annual production averaged over the first three years, and the cost curve for unit production is fit to the entire 20-year series. The production curve equation in this case is displayed in Figure D.4. If an existing system that we use to project the units procured for the ATO-derived system has only, say, ten years of production history, the full 20-year production profile can be projected based on the curve.

Figure D.4
Annual Purchases of an Electronics System over a 20-Year Period

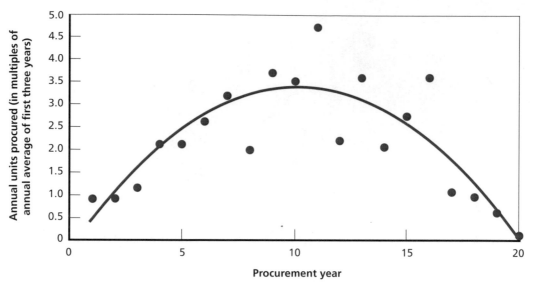

NOTES: $y = -0.0346x^2 + 0.7097x - 0.2804$. $R^2 = 0.7092$.

   This figure represents an introductory rate of a new support equipment system (based on existing systems still being acquired in 2006).

RAND *MG761-D.4*

From the production equation, the first year of production for the average electronics system accounts for a normalized level of production of 0.39, which is derived from $(-0.0346(1)^2 + 0.7097(1) - 0.2804)$. The total area under the production curve is 44.13. These values in and of themselves are meaningless. Rather, they indicate that historically, the first year of production is 0.89 percent of the full 20-year production schedule ($0.39 \div 44.13$). We can calculate a similar percentage for each year and add these numbers together to suggest the percentage of production that has already taken place at any point under the production curve. If ten years of production have already passed, this represents 51.53 percent of total production. Therefore, if 100 units of a weapon system have been produced over a ten-year period, we project that approximately 194 (i.e., $100 \div 0.5153$) units will be produced over a full 20-year period.

If the weapon system is vehicular, the projection is based on a corresponding universal curve for historical support equipment production (see Figure D.5). To develop this curve, we used 11 existing support equipment systems, which consist mainly of vehicular systems, such as the M-1097 High Mobility Multipurpose Wheeled Vehicle and the families of heavy and medium tactical vehicles.

Figure D.5
Vehicular System Production As a Ratio of First Three Years

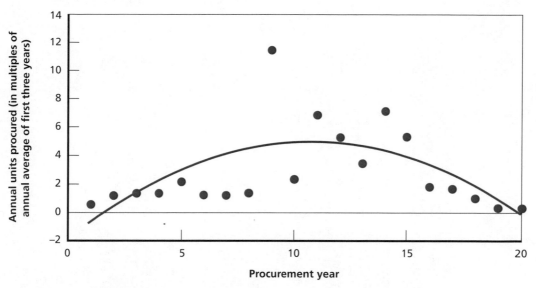

NOTES: y = −0.0589x^2 + 1.2742x − 2.0508. R^2 = 0.3875.

This figure represents an introductory rate of a new support equipment system (based on existing systems still being acquired in 2006).

Choosing Essential ATOs Across a Wide Range of Capability Requirements

The methodology for ranking projects across a range of future requirements can be easily applied to a much wider range of uncertainties. Figure E.1 displays selection results for each project when expected requirements fluctuate by up to 50 percent from the reference requirement, as opposed to 20 percent in the main text. Five total levels are shown. This expands the range shown in Figure 5.3.

When capabilities are increased to 150 percent of the reference requirement (1.5 × RR), 16 of the 29 projects are selected. At 50 percent of the reference requirement, only eight projects are needed to meet the reduced requirement. The portfolio manager can expect much more variation in terms of the mix and number of projects required across levels when the requirement range is increased. Table E.1 confirms these expectations. Compared with the results in Table 5.2, only 12 projects are never selected here, as opposed to 14 previously. Fewer projects are selected five times. Moreover, each cell of the table is filled, so that some ATOs are selected from one up to four times across five requirement levels.

Determining which ATOs are needed to meet a range of requirements remains simple. As shown in Table 5.2, 15 ATOs are needed to meet the future requirement, which turns out to be somewhere within plus and minus 20 percent from the reference requirement. Here, we show that if the uncertainty in future requirements is larger (50 percent instead of 20 percent), the Army would need to have more ATOs, 17, as shown in Table E.1. Our model shows how many and which ATOs are needed under a given range of future requirements.

Figure E.1
ATOs That Meet Various FOC Requirements at Lowest Remaining Lifecycle Cost

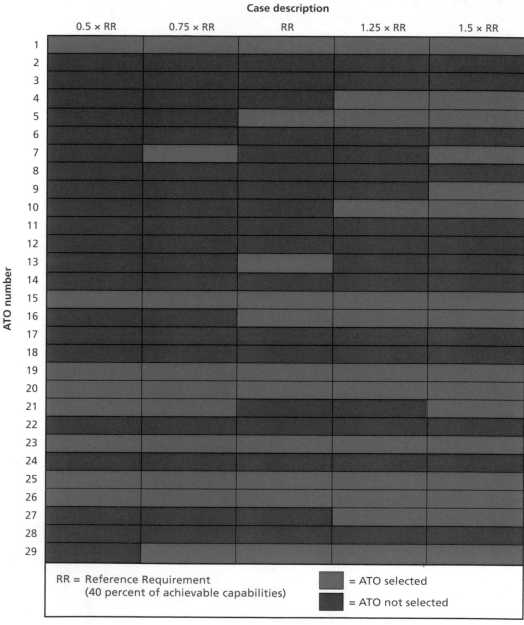

Table E.1
Selection of ATOs Across Requirement Levels
(within 50 percent of the reference requirement)

Selection Count	ATO Number (1 through 29)
5	1, 15, 19, 20, 23, 25, 26
4	29
3	5, 16, 21
2	4, 7, 10, 27
1	9, 13
0	2, 3, 6, 8, 11, 12, 14, 17, 18, 22, 24, 28

Insensitivity of Model Results to Detailed Estimation of Remaining Lifecycle Cost

In the approach discussed in the main text, marginal remaining lifecycle cost (MRLCC) was estimated in dollar terms for each ATO. We call this approach *the dollar approach.* These MRLCCs for the 29 ATOs can differ widely. This appendix examines how sensitive the model results are to the precise values of these MRLCCs. An interesting question is whether rough MRLCCs expressed in several grade levels, as opposed to precise dollar values that required much more effort to estimate, would yield similar results. If so, one may not need to go through the considerable efforts in estimating precise cost, and a Delphi method in estimating these MRLCC grade levels may be a useful tool when detailed cost estimates are unavailable. We call this approach *the grade approach.* Moreover, the remaining lifecycle cost is rather uncertain at the ATO stage; therefore, a grade approach may suffice.

For *the grade approach*, we assign each ATO's marginal implementation cost a numerical grade from –2 to 4. The marginal implementation cost (MIC) is defined as the cost over the legacy system to develop, demonstrate, acquire, and operate the new systems after an ATO is completed. The MIC is equal to MRLCC minus RSTC. Both MRLCC and RSTC can be found in Table 4.2. A numerical grade is assigned to each ATO's MIC according to Table F.1. If an ATO can develop systems less expensively than the legacy systems to be replaced and yield a savings up to $300 million, this ATO's MIC is assigned a grade of –1. For all cases in which MIC numbers were assigned as negative, the pertaining DTO data sheet would have stated or implied a savings in MIC. For example, the procurement cost of the new system is expected to be below that of the legacy system. Another example is that the new system is specifically designed to reduce O&M costs relative to the legacy system. Since reducing implementation cost is an objective for the ATO, we want to capture that positive attribute by crediting the ATO's MIC with a negative grade (i.e., cost savings). On the other hand, if an ATO's systems are more expensive than the legacy systems and, say, result in an increase in MIC of up to $150 million, the ATO's MIC is assigned a grade of 0. Other assignments are made according to the scale used in Table F.1.

Table F.1
Numerical Grade Description for Implementation Cost

Grade	Description	ATOs in Group
–2	Cost decrease of more than $300 million	0
–1	Cost decrease from more than $0 million to $300 million	2
0	Zero to $150 million increase	11
1	Cost increase from more than $150 million to $300 million	4
2	Cost increase from more than $300 million to $2 billion	6
3	Cost increase from more than $2 billion to $4 billion	2
4	Cost increase above $4 billion	4

The third column of Table F.1 displays the number of ATOs whose MICs fall into each numerical grade. In order to compare the model results from this grade approach with *the dollar approach*, we need to add back the RSTC to the graded implementation cost in order to arrive at an MRLCC for the grade approach. We use four examples to show how we translate the RSTC and MRLCC to a numerical grade. The first example assumes that the ATO has an MIC of 1 and an RSTC of $30 million. We use the lower bound of grade 1, $150 million, as the baseline. The $30 million cost is equivalent to a numerical grade of 0.2, because the range for grade 1 is $150 million (i.e., $300 million – $150 million), and the $30 million cost brings an additional 0.2 (i.e., 30 ÷ 150) above grade 1 to yield an MRLCC of 1.2. A second example assumes that the MIC is –1 and the RSTC is $400 million. Then, the baseline is again the lower bound of grade –1 or negative $300 million. However, the $400 million cost brings the MRLCC to grade 0 and, in fact, two-thirds of the way toward grade 1 (i.e., (400 – 300) ÷ 150). Thus, MRLCC is 0.67. A third example looks at a case that is an exception to the above rule. The MIC is –2, and the RSTC is $60 million. If we followed the same rule of using the lower limit, we would use negative infinity, making a grade assignment for MRLCC infeasible. To alleviate this problem for grade –2, we use a lower limit that is its upper limit minus $300 million; the latter is the range width for grade –1. Then, the lower limit for grade –2 is –$300 million – $300 million = –$600 million. Since the RSTC of $60 million is 0.2 of the range width for grade –2, it brings the MRLCC 0.2 above the grade of –2, or an MRLCC of –1.8. A fourth example is another similar exception for an MIC in the highest grade of 4. The RSTC is $60 million. As for the lowest grade, we assume that the range width for grade 4 is the same as the next highest grade of $2 billion. Then the RSTC adds 0.03 (i.e., 60 ÷ 2,000) to 4 to yield an MRLCC of 4.03.

These assigned numerical grades were substituted for the dollar MRLCC estimations in the linear programming model. The ATOs selected by the model for providing the lowest total remaining lifecycle cost using the *grade* approach are compared with those selected according to the *dollar* approach in Figure F.1. As in Figure 5.6, green cells indicate projects that are selected, while red cells indicate projects that are rejected for the optimal ATO portfolios at specific requirement levels. At all requirement levels, the results between the *dollar* and *grade* approaches are exactly the same. Thus, in this example, rough cost estimates as long as they are in the right level are adequate. The *dollar* approach suggests that the Army should protect 14 projects (1, 5, 15, 16, 19, 20, 23, 25, 26, 29, 10, 4, 13, 21, and 27) in order to produce the lowest total remaining lifecycle cost over the range of future requirement from 0.8RR to 1.2RR as indicated in Figure 5.6 and Table 5.2. At the same time, the grade approach would suggest exactly the same set of ATOs to be protected. This result indicates that a Delphi method can potentially play a useful role in the methodology proposed in this study. Since the Delphi method is a much simpler technique than the cost estimation methods described in this monograph and it can make our analytic framework much easier to implement, we believe that it warrants serious consideration.

Figure F.1
ATOs That Meet Various FOC Requirements at Lowest Remaining Lifecycle Cost (dollar versus grade)

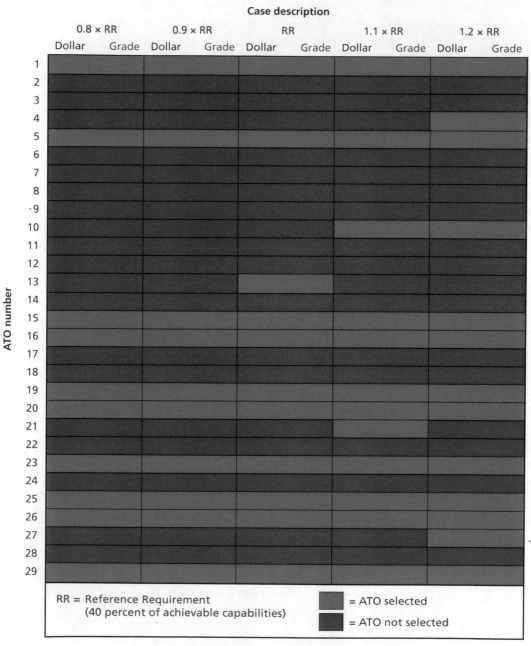

Collate Color Text Section

Collate Color Text Section

Bibliography

Anderton, Gary L., "Rapid Contingency Airfields," *The Military Engineer*, October 2005, pp. 97–637.

Andrews, Richard, *An Overview of Acquisition Logistics*, Kettering, Ohio: Air Force Institute of Technology, School of Systems and Logistics, undated. As of January 23, 2009:
https://acc.dau.mil/GetAttachment.aspx?id=142351&pname=file&aid= 27644&lang=en-US-

"Armored Medical Evacuation Vehicle AMEV," *GlobalSecurity.org*, last modified April 27, 2005. As of January 23, 2009:
http://www.globalsecurity.org/military/systems/ground/amev.htm

Army Logistics Management College, *Analysis of Alternatives*, Fort Lee, Va.: Army Logistics Management College, PM-2009-DL, June 7, 2005.

Bennis, Phyllis, "A Failed 'Transition': The Mounting Costs of the Iraq War," Washington, D.C.: Institute for Policy Studies, September 30, 2004.

"Boeing Project Takes Aim at Wars of the Future," *KOMONews.com*, undated. As of January 23, 2009:
http://www.komonews.com/news/archive/4157681.html

Burges, Lisa, "U.S. Troop Presence in Afghanistan at 17,900 and Expected to Hold Steady," July 9, 2004. As of January 23, 2009:
http://www.globalsecurity.org/military/library/news/2004/07/mil-040709-arng01.htm

Center for Program Management, *New Acquisition Policy, 12 May 2003*, Fort Belvoir, Va.: Defense Acquisition University, May 19, 2003.

Chairman of the Joint Chiefs of Staff (CJCS), *Joint Capabilities Integration and Development System*, Fort Belvoir, Va.: Defense Technical Information Center, CJCSI 3170.01F, May 1, 2007a.

———, *Operation of the Joint Capabilities Integration and Development System*, Fort Belvoir, Va.: Defense Technical Information Center, CJCSM 3170.01C, May 1, 2007b.

Christensen, Bill, "WeaponWatch Detects, Locates Enemy Fire IR Signature," *Technovelgy*, October 18, 2006. As of January 23, 2009:
http://www.technovelgy.com/ct/Science-Fiction-News.asp?NewsNum=774

CJCS—*see* Chairman of the Joint Chiefs of Staff.

Damstetter, Donald, "Life Cycle Management Improvement Initiatives," Office of the Deputy Assistant Secretary of the Army for Plans, Programs, and Resources (DASA [PPR]) and Office of the Assistant Secretary of the Army for Acquisition, Logistics and Technology (OASA [ALT]), November 16, 2004.

DDR&E—*see* Director of Defense Research and Engineering.

Director of Defense Research and Engineering (DDR&E), *Defense Technology Objectives for the Joint Warfighting Science and Technology Plan (JWSTP)*, Arlington, Va., February 2006.

Edwards, Travis, "First MQ-9 Reaper Makes Its Home on Nevada Flightline," *GlobalSecurity.org*, March 14, 2007. As of January 23, 2009:
http://www.globalsecurity.org/military/library/news/2007/03/mil-070314-afpn07.htm

Greenberg, Marc, and James Gates, *Analysis of Alternatives*, Fort Belvoir, Va.: Defense Acquisition University, April 2006.

Haynes, Mary L., *Department of the Army Historical Summary, Fiscal Year 1987*, Washington, D.C.: Center of Military History, U.S. Army, 1995. As of January 23, 2009:
http://www.history.army.mil/books/DAHSUM/1987/index.htm

Hoffman, Natalie, "Napa Gets State's First Hybrid School Bus," *NapaValleyRegister.com*, August 10, 2007. As of February 20, 2009:
http://www.napavalleyregister.com/articles/2007/08/10/news/local/doc46bcca6521e51974045742.txt

"Hybrid Technology," *IC Corporation*, Warrenville, Ill.: IC Bus, LLC, undated. As of January 23, 2009:
http://www.ic-corp.com/Static%20Files/ICCorp/PageContent/Community/Hybrid%20Power/Hybrid%20FAQs.pdf

icasualties.org, Iraq Coalition Casualty Count, Iraq, Afghanistan, "U.S. Wounded, by Month," undated. As of January 23, 2009:
http://icasualties.org/oif/

"Lockheed Wins $149.2M Contract for High Altitude Airship," *Defense Industry Daily*, January 16, 2006. As of January 23, 2009:
http://www.defenseindustrydaily.com

"Manned Ground Vehicle," *GlobalSecurity.org*, January 2005. As of January 23, 2009:
http://www.globalsecurity.org/military/systems/ground/fcs-mgv.htm

Michaels, Jim, "Attacks Rise on Supply Convoys: Civilian Guards Provide Security," *USA Today*, July 9, 2007.

Moteff, John, *Defense Research: DoD's Research, Development Test and Evaluation Program*, Washington, D.C.: Congressional Research Service, updated March 10, 2003.

Moulder, Roger, *Engineering/Integrated Product and Process Development (IPPD) in Science & Technology*, Pickerington, Ohio: James Gregory Associates, Inc., undated.

"MQ-9A Predator B," GlobalSecurity.org, last modified June 10, 2007. As of January 23, 2009:
http://www.globalsecurity.org/intell/systems/predatorb.htm

Nicholas, Ted, and Rita Rossi, *U.S. Weapon Systems Costs, 2002*, Fountain Valley, Calif.: Data Search Associates, 2002.

———, *U.S. Weapon Systems Costs, 2006*, 26th ed., Fountain Valley, Calif.: Data Search Associates, April 2006.

———, *U.S. Weapon Systems Costs, 2007*, Fountain Valley, Calif.: Data Search Associates, 2007.

———, *U.S. Weapon Systems Costs, 2008*, Fountain Valley, Calif.: Data Search Associates, 2008.

Office of Science and Technology Policy, *Discovery, Education and Innovation, An Overview of the Federal Investment in Science & Technology,* undated. As of January 23, 2009:
http://www.ostp.gov/galleries/NSTC%20Reports/Discovery%20Education%20Innovation%20 2000.pdf

"Pioneer Short Range (SR) UAV," *GlobalSecurity.org,* last modified April 26, 2005. As of January 23, 2009:
http://www.globalsecurity.org/intell/systems/pioneer.htm

"RQ-1 Predator MAE UAV," GlobalSecurity.org, last modified May 14, 2008. As of January 23, 2009:
http://www.globalsecurity.org/intell/systems/predator.htm

"RQ-4A Global Hawk (Tier II+ HAE UAV)," GlobalSecurity.org, last modified May 14, 2008. As of January 23, 2009:
http://www.globalsecurity.org/intell/systems/global_hawk.htm

Shalal-Esa, Andrea, "Pentagon Trims Armored Vehicles Due in '07 for Iraq," Reuters, July 19, 2007. As of January 23, 2009:
http://www.reuters.com/article/idUSN1939330420070720

Shepard, Donald S., "Cost-Effectiveness of a Pediatric Dengue Vaccine," *Vaccine,* Vol. 22, March 12, 2004, pp. 9–10.

Silberglitt, Richard, and Lance Sherry, *A Decision Framework for Prioritizing Industrial Materials Research and Development,* Santa Monica, Calif.: RAND Corporation, MR-1558-NREL, 2002. As of January 23, 2009:
http://www.rand.org/pubs/monograph_reports/MR1558/

Silberglitt, Richard, Lance Sherry, Carolyn Wong, Michael Tseng, Emile Ettedgui, Aaron Watts, and Geoffrey Stothard, *Portfolio Analysis and Management for Naval Research and Development,* Santa Monica, Calif.: RAND Corporation, MG-271-NAVY, 2004. As of January 23, 2009:
http://www.rand.org/pubs/monographs/MG271/

Smith, Roger, "Serving 1,000,000 Global Customers: How Can We Offer Training Anywhere Anytime?" U.S. Army, Program Executive Office for Simulation, Training & Instrumentation, briefing presented at the Joint Advanced Distributed Learning Fest, Orlando, Fla., August 28, 2007. As of January 23, 2009:
http://www.peostri.army.mil/CTO/FILES/OneMillionCustomers.pdf

"Surviving a Tour of Duty," *The Abstract Factory,* June 26, 2005. As of January 23, 2009:
http://www.abstractfactory.blogspot.com/2005/06/surviving-tour-of-duty.html

Touchette, Nancy, "New Approaches Speed Ebola Vaccine Development," *Genome News Network,* March 19, 2004. As of January 23, 2009:
http://www.genomenewsnetwork.org

U.S. Air Force, *C130 Hercules,* fact sheet, Scott Air Force Base, Ill.: Air Mobility Command, September 2008. As of January 23, 2009:
http://www.af.mil/factsheets/factsheet.asp?fsID=92

U.S. Army, *1998 Army Science and Technology Master Plan,* "Science and Technology Objectives," Arlington, Va.: Assistant Secretary of the Army, Research, Development, and Acquisition, March 1998, Figures I-11 and I-12. As of January 23, 2009:
http://www.fas.org/man/dod-101/army/docs/astmp98/sec1c.htm

———, *Army Acquisition Procedures,* Arlington, Va.: Department of the Army, Headquarters, Pamphlet 70-3, July 15, 1999.

————, *Army Acquisition Policy*, Arlington, Va.: Department of the Army, Headquarters, Army Regulation 70-1, December 31, 2003.

————, Assistant Secretary of the Army for Acquisition, Logistics and Technology, *2005 United States Army Weapon Systems*, March 23, 2005a.

————, *Military Operations: Force Operating Capabilities*, Fort Monroe, Va.: Training and Doctrine Command, pamphlet 525-66, July 1, 2005b.

————, Assistant Secretary of the Army for Acquisition, Logistics and Technology, *Army Science & Technology Master Plan*, Volume II—Annexes, July 2005c.

————, Deputy Assistant Secretary of the Army, *Army Science and Technology Master Plan*, Volume 1, July 2005d.

————, *2006 Army Modernization Plan*, Washington, D.C.: U.S. Army Deputy Chief of Staff, G-8, March 2006. As of January 23, 2009:
http://www.army.mil/features/MODPLAN/2006/

————, *Descriptive Summaries for Program Elements of the Research, Development, Test and Evaluation, Army FY 2009 Budget Estimate*, Volume II, Arlington, Va.: Department of the Army, Office of the Secretary of the Army (Financial Management and Comptroller), February 2008.

U.S. Army Corps of Engineers, Joint Rapid Airfield Construction, 2007 Demonstration Project, "Project Spotlight," undated. As of January 23, 2009:
https://jrac.erdc.usace.army.mil/jrac.html

U.S. Army, Cost and Economic Analysis Center, Operating and Support Management Information System (OSMIS), "Business Story," Arlington, Va.: Department of the Army, Headquarters, undated. As of February 23, 2009:
https://www.osmisweb.army.mil/osmisrdb/unsecure/WhatIsOsmis.aspx

U.S. Department of Defense, "Multi-Analyte, Wearable Chemical Nanosensor for Warfighter Physiological Status Monitor (WPSM)," SBIR/STTR Interactive Topic Information System (SITIS), undated. As of February 17, 2009:
http://www.dodsbir.net/SITIS/archives_display_topic.asp?Bookmark=30467

————, Office of the Inspector General, *Acquisition: Army Transition of Advanced Technology Programs to Military Applications (D-2002-107)*, June 14, 2002.

————, Under Secretary of Defense for Acquisition, Technology and Logistics (USD [AT&L]), *Operation of the Defense Acquisition System*, Department of Defense Instruction 5000.1, May 12, 2003a.

————, Under Secretary of Defense for Acquisition, Technology and Logistics (USD [AT&L]), *Operation of the Defense Acquisition System*, Department of Defense Instruction 5000.2, May 12, 2003b.

————, *National Defense Budget Estimates for FY 2008*, Washington, D.C.: Office of the Under Secretary of Defense (Comptroller), March 2007. As of January 23, 2009:
http://www.defenselink.mil/comptroller/defbudget/fy2008/fy2008_greenbook.pdf

"U.S. Forces Order of Battle," *GlobalSecurity.org*, last modified May 19, 2008. As of January 23, 2009:
http://www.globalsecurity.org/military/ops/iraq_orbat.htm

U.S. Government Accountability Office, Best Practice: *Stronger Practices Needed to Improve DoD Technology Transition Processes*, Washington, D.C.: U.S. Government Accountability Office, GAO-06-883, September 14, 2006.

The Value Line Investment Survey, No. 9, October 26, 2007.

Vasquez, Susie, "Valley Soldier Facing Tough Choice," *Record-Courier*, April 27, 2005. As of January 23, 2009:
http://www.recordcourier.com/article/20050427/News/104270018/-1/NEWS/

Yenne, Bill, *Attack of the Drones: A History of Unmanned Aerial Combat*, Osceola, Wisc.: Zenith Press, 2004.

"Young: Pentagon Should Invest Billions More in Science, Technology," *Inside the Pentagon*, September 13, 2007.